FOREWORD

Writing a children's book is something I've dreamed of since my own school days. Since it is becoming more difficult in our world to access clean water, I felt it was time that I made that dream a reality – for myself, for my darlings, Sue and Wolf, and for anyone who wants to know more about that intriguing substance, water. I would very much like to thank our publisher, Sofie Van Sande of Uitgeverij Lannoo, for making this book possible, as well as Marijke Huysmans and Wendy Panders for cowriting and illustrating it. I hope you will find what you're looking for here, and I hope it will inspire you to dive even deeper into the wondrous world of water!

Sarah Garré

SARAH GARRÉ AND MARIJKE HUYSMANS
WITH ILLUSTRATIONS BY WENDY PANDERS

THE WONDERFUL WORLD OF WATER

FROM DAMS TO DESERTS

PRESTEL
Munich · London · New York

8
WATER BRINGS LIFE
10
THE BLUE PLANET
12
SEAS AND OCEANS
14
WHAT IS WATER?
16
THE WATER CYCLE
18
WATER EVERY-WHERE
20
WHERE DOES GROUND-WATER COME FROM?
22
HOW DO YOU FIND GROUNDWATER?
24
THE ROLE OF WATER IN THE CLIMATE
26
WATER AND WEATHER
28
HOW DID PEOPLE DEAL WITH WATER IN THE PAST?
30
WHY DO YOU NEED TO DRINK WATER?
32
HOW DOES WATER PURIFICATION ACTUALLY WORK?

34
HOW DOES DRINKING WATER REACH YOUR HOME?
36
WHAT HAPPENS TO YOUR URINE?
38
WITHOUT WATER, YOUR PLATE WOULD BE EMPTY
40
WHAT IS VIRTUAL WATER?
42
HOW DO PLANTS TAKE WATER OUT OF THE GROUND?
44
LIFE IN THE RAIN FOREST
46
SURPRISES IN THE DESERT: OASES
48
WHAT IS TIGER BUSH?
50
RIVERS AND FLOODS
52
WAVES AND TSUNAMIS
54
DAMS AND RESER-VOIRS
56
THE FUTURE OF WATER IN THE WORLD
58
HOW CAN YOU SAVE WATER?

WATER BRINGS LIFE

Do you love summer as much as we do? What does "good weather" mean to you? Many people like it when it doesn't rain for a long time, when it's warm and sunny. You can go for a swim with your friends, or head to the beach, or perhaps even enjoy some ice cream! All the same, we really do need rain! Drought can cause big problems. Historians believe that water shortages may even have played a role in the fall of some major civilizations, such as the Mayan civilization in Mesoamerica. Water is very important to everyone in our society.

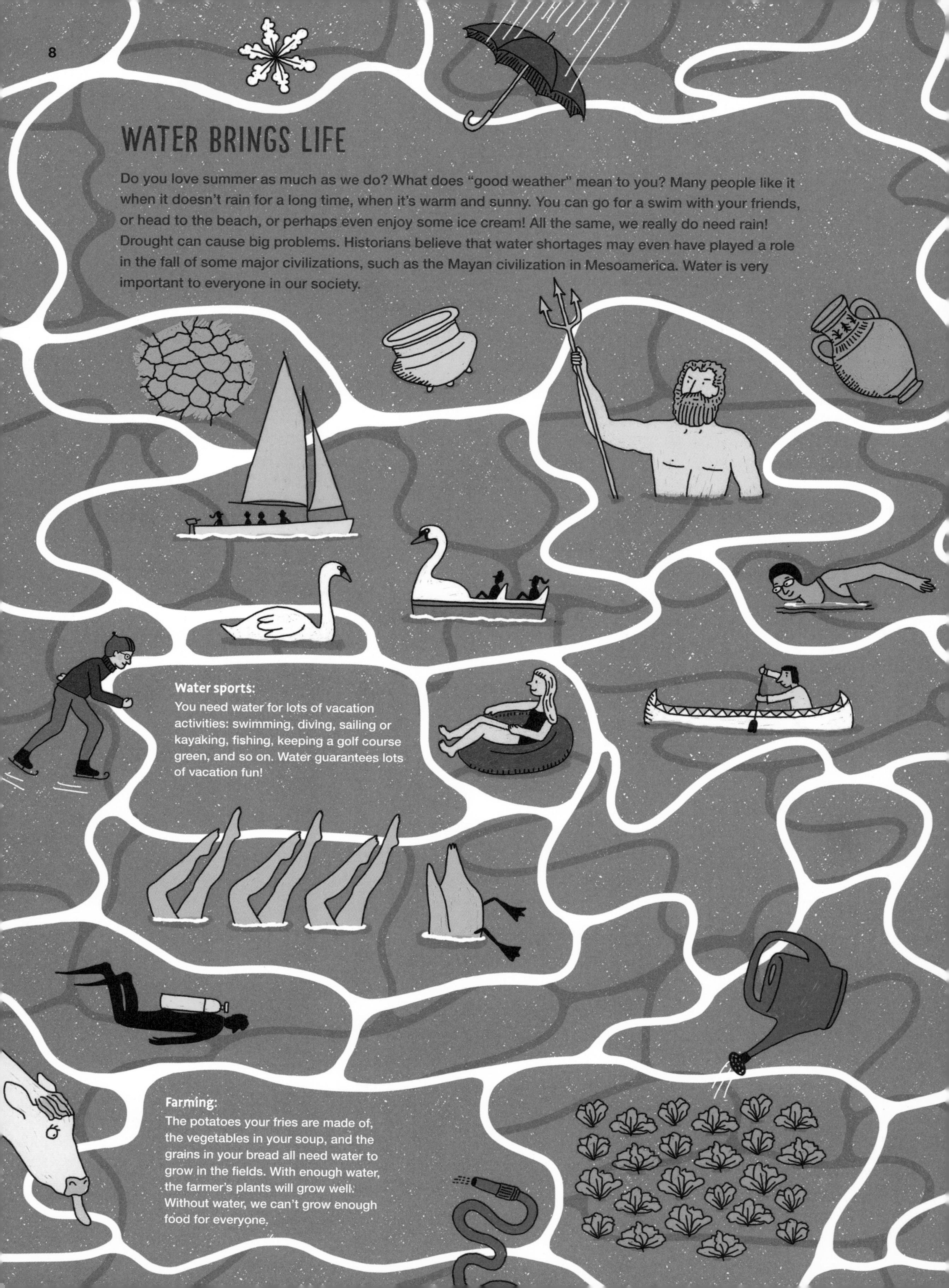

Water sports:
You need water for lots of vacation activities: swimming, diving, sailing or kayaking, fishing, keeping a golf course green, and so on. Water guarantees lots of vacation fun!

Farming:
The potatoes your fries are made of, the vegetables in your soup, and the grains in your bread all need water to grow in the fields. With enough water, the farmer's plants will grow well. Without water, we can't grow enough food for everyone.

Factories:

In order to work safely, the machines in factories regularly need to be cooled down with water.

Nature:

Plants, trees, and flowers need water to grow. You've probably noticed that the grass turns brown when it doesn't rain for a while. Fortunately, it goes green again as soon as more rain comes. But there are many plants that don't deal so well with drought. Fish and other aquatic animals also need enough water in the rivers.

Transport:

Ships are ideal for transporting big, heavy items. They ensure that our factories have enough sand, cement, chemicals, and fuel to manufacture the things we want and need. Once in a while, there needs to be enough rain so that the water level in the rivers and canals doesn't drop too low.

At home:

We also use lots of water in our homes. When it's dry outside, we have to use faucet water sparingly, so we shouldn't use it to wash the car, water the yard, or fill the swimming pool.

THE BLUE PLANET

In the morning, you turn on the faucet to wash, and there it is: water. It seems so obvious, but it's actually very special. Three-fourths of Earth's surface consists of water. Because there's so much water, astronauts see Earth as a blue planet from their spaceship. But oceans are not really deep at all compared with the size of Earth. There is only a thin layer of water on the surface. So even on the blue planet, water is precious. Earth floats 93 million miles / 150 million kilometers from the sun. At that distance, our planet is not too warm and not too cold. That's how oceans and seas can exist. Close to the sun, the water immediately evaporates and disappears. Farther away, it's so cold that water can exist only as ice. No one knows for certain, but many scientists think that without liquid water, there could be no life on Earth.

Is our Earth the only heavenly body with water?

No. Ice has been found on Earth's moon, and traces of water have also been discovered on Mars. Some moons around Jupiter and Saturn even have very deep oceans under a thick layer of ice.

How much water is there on Earth?

More than 326 million trillion gallons / a sextillion liters of water.

13.8 BILLION YEARS AGO
The Big Bang:
The Universe begins.

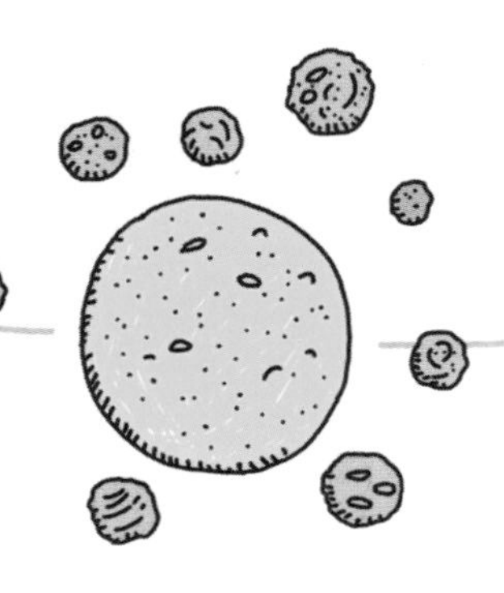

4.54 BILLION YEARS AGO
Earth forms but doesn't have any seas yet.

Comets and planetoids with water bombard Earth.

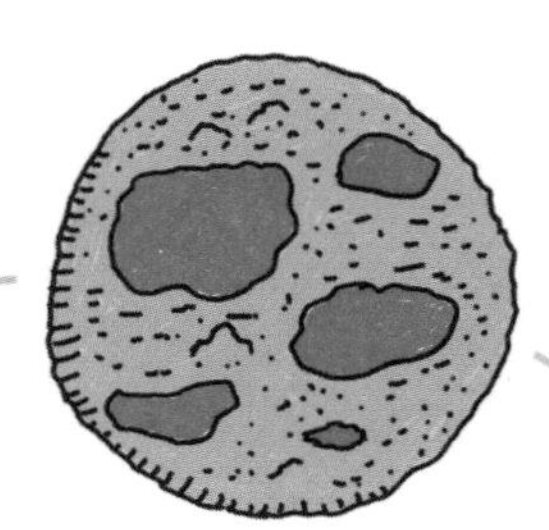

3.8 BILLION YEARS AGO
Oceans appear on Earth but not where they are now.

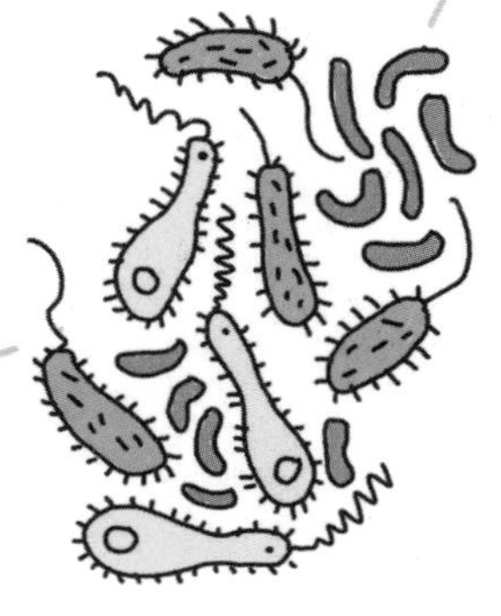

3.7 BILLION YEARS AGO
Life begins on Earth. For a very long time, the only inhabitants are bacteria.

Between 1.1 billion years ago and 900 million years ago, all the continents together form the supercontinent Rodinia.

650 MILLION YEARS AGO
The oceans freeze, and the land is covered in ice.

500 MILLION YEARS AGO
More and more animal species conquer the oceans.

They live in the water and on the land.

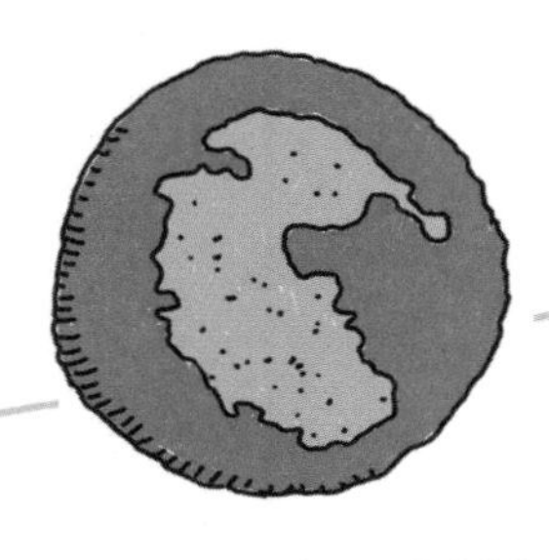

299 MILLION YEARS AGO
Pangea, a new supercontinent, is formed.

The bare earth turns greener. Plants have started growing on the land.

252 MILLION YEARS AGO
Dinosaurs populate Earth.

66 MILLION YEARS AGO
The dinosaurs die out.

During the ice ages, large ice caps form on Earth.

The mammoth is having a great time.

25,000 years ago, the last ice age ends.

Where does all that water on Earth come from?

Before life began on Earth, millions of enormous blocks of ice hit the planet. Scientists think that this was the origin of almost all the water on Earth, but they're not entirely certain.

SEAS AND OCEANS

When you think of water, do you think of salty seawater? Compared with the water in the seas and oceans, there is relatively little freshwater in lakes, rivers, glaciers, and rainstorms. Under the waves of the sea is a strange, wild world of high mountains and impressive valleys. All kinds of plants and animals live there, like on land. The animals can be very small or very big. The biggest animal on Earth is the blue whale, which can grow up to 100 feet / 30 meters long. That's as long as two buses, one behind the other!

blue whale

There are five oceans on Earth and they're all connected: the Pacific Ocean, the Atlantic Ocean, the Indian Ocean, the Southern Ocean, and the Arctic Ocean. Oceans are so deep and dark that we know very little about them. Scientists say that we have explored only one-twentieth of all there is to see. It's crazy, really, that today we know more about space and other planets than about our own seas and oceans!

hydrothermal vents

vampire squid

carbonated chimneys of the Lost City

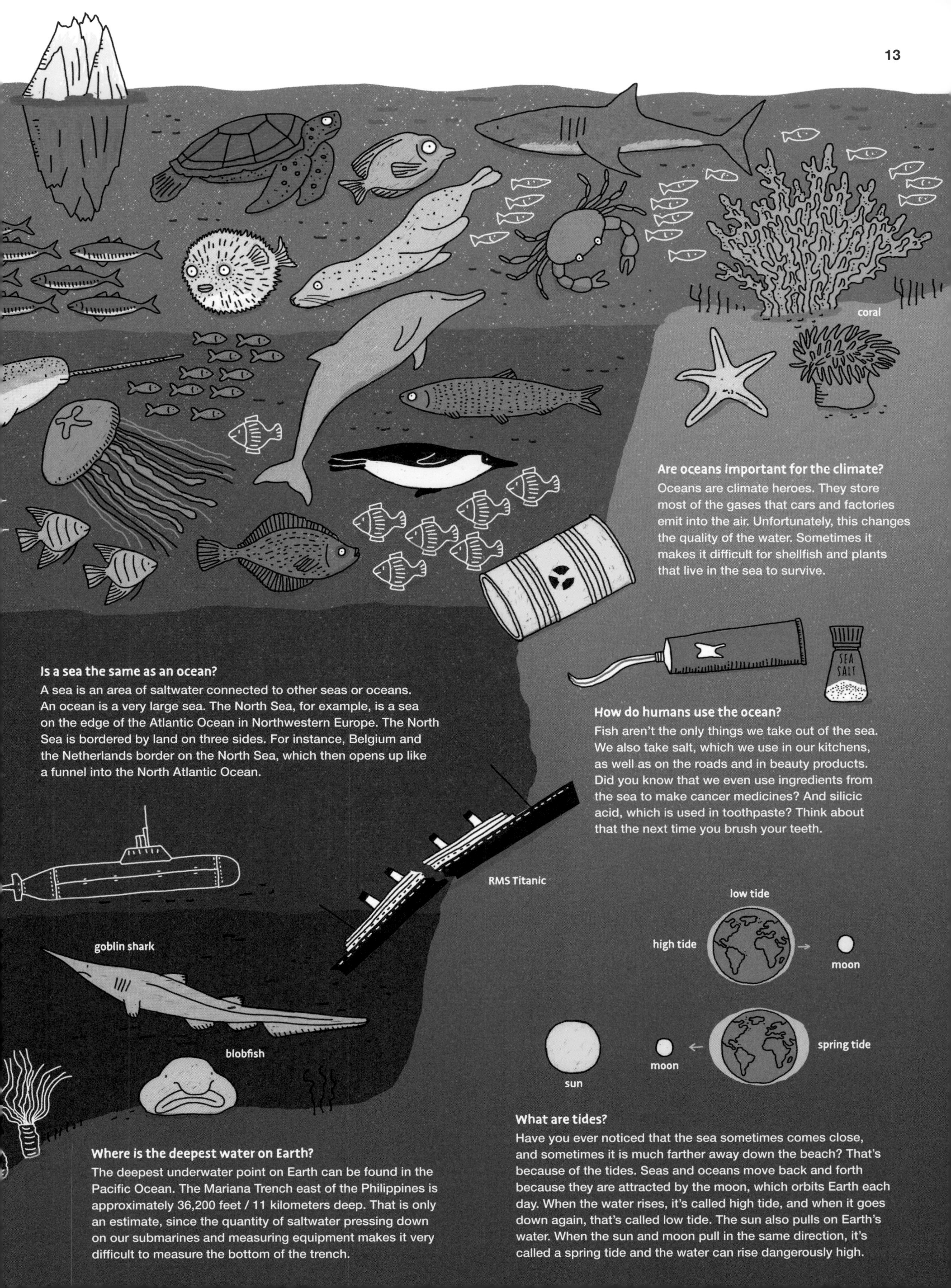

Are oceans important for the climate?

Oceans are climate heroes. They store most of the gases that cars and factories emit into the air. Unfortunately, this changes the quality of the water. Sometimes it makes it difficult for shellfish and plants that live in the sea to survive.

Is a sea the same as an ocean?

A sea is an area of saltwater connected to other seas or oceans. An ocean is a very large sea. The North Sea, for example, is a sea on the edge of the Atlantic Ocean in Northwestern Europe. The North Sea is bordered by land on three sides. For instance, Belgium and the Netherlands border on the North Sea, which then opens up like a funnel into the North Atlantic Ocean.

How do humans use the ocean?

Fish aren't the only things we take out of the sea. We also take salt, which we use in our kitchens, as well as on the roads and in beauty products. Did you know that we even use ingredients from the sea to make cancer medicines? And silicic acid, which is used in toothpaste? Think about that the next time you brush your teeth.

What are tides?

Have you ever noticed that the sea sometimes comes close, and sometimes it is much farther away down the beach? That's because of the tides. Seas and oceans move back and forth because they are attracted by the moon, which orbits Earth each day. When the water rises, it's called high tide, and when it goes down again, that's called low tide. The sun also pulls on Earth's water. When the sun and moon pull in the same direction, it's called a spring tide and the water can rise dangerously high.

Where is the deepest water on Earth?

The deepest underwater point on Earth can be found in the Pacific Ocean. The Mariana Trench east of the Philippines is approximately 36,200 feet / 11 kilometers deep. That is only an estimate, since the quantity of saltwater pressing down on our submarines and measuring equipment makes it very difficult to measure the bottom of the trench.

WHAT IS WATER?

The water in your glass has no color or odor of its own. When you place it in the freezer, it turns to ice. When you boil it on the stove, it turns to gas or steam. But what is it precisely? Well, it's a structure of very small particles. We call the structure a molecule. The water molecule consists of two types of atoms: one oxygen atom (O) and two smaller hydrogen atoms (H). That's why scientists call water H_2O. That's logical, right?

Water can take on many different forms. When it gets colder than 32°F / 0°C, it changes into ice. That's when you need to watch out for cracks, since water takes up more space when it's frozen than when it is liquid. If you put a sealed bottle of water in the freezer, it can explode. When water gets warmer than 212°F / 100°C, it begins to boil. Bubbles appear on the surface because the water molecules change one by one into steam and hang invisibly in the air. Ice can also change directly into steam without first becoming water. We call that sublimation.

Why does water boil faster up a mountain?
Water boils much faster than normal when you're up a mountain, or in fact, it boils at a lower temperature. So your potatoes won't soften as fast. Here's why: The boiling point of water at high altitude is a bit lower because the air pressure is lower than at sea level. The air pressure tells us the force of the air pushing down on us and everything around us. That force decreases as you go higher in the air. In the Alps, for instance, water already boils at 194°F / 90°C.

Why isn't seawater transparent?
The sea often seems to have a blue color. That's because of the light on Earth. The light is white, and it contains lots of different colors. Water absorbs red light 100 times more than blue light. Water also scatters blue light five times more than red light. The sea looks blue to our eyes because water absorbs red light and reflects blue light. This effect becomes clear only when the water is deeper than 15 feet / 5 meters.

Can you make seawater drinkable?
In very dry, hot countries, such as Saudi Arabia, there are enormous factories that supply drinking water to people living in the desert. The seawater is heated up in big desalination factories to produce steam. The salt is left behind, and then in a different area, the steam condenses to liquid again as drinking water. Unfortunately, it takes a lot of energy to heat the water.

FRESH-WATER
SPARK-LING WATER
SWIMMING POOL WATER
HOLY WATER
FIRE-FIGHTING WATER
STRONG WATER (ACID)
STILL WATER
SOFT WATER
HARD WATER
SURFACE WATER
GROUND-WATER
BRACKISH WATER
condensation
evaporation
vapor
sublimation
DITCH-WATER
melting
freezing
H2O
RAIN-WATER
DRAIN WATER
FAUCET WATER

THE WATER CYCLE

You certainly have felt fresh raindrops on your face at some point, but do you actually know where that rain comes from? Those raindrops aren't new water. That water has always been there, and it travels round and round on an endless journey known as the water cycle.

Water in oceans, seas, lakes, and ponds on the ground evaporates due to the warmth of the sun. The water vapor rises into the air and forms a cloud when it reaches higher, colder layers of air. When lots of water droplets have clumped together to form a cloud, rain, hail, or snow can eventually fall from it.

Many of the raindrops land back in the seas and oceans, but of course, some also rain on land. What happens to the rainwater when it reaches the ground? Most of the raindrops in and on the ground evaporate again in the warmth of the sun. The warmer it is, the more water evaporates.

Plants also drink lots of the rainwater. Much of the water their roots draw out of the ground goes back into the air through the plant's leaves. We call that transpiration. In fact, plants sweat, just like humans.

Some of the rainwater runs across the land and into streams and rivers, which eventually carry the water to the sea.

Some of the rainwater drains into the ground. That is called infiltration. Infiltration is very important for groundwater, the water located under our feet in the ground. The groundwater flows very slowly, but that water, too, finally ends up in the rivers and the sea.

Once in the sea, the water can evaporate again and the water cycle begins again. And again. And again.

Why is seawater salty?

Step by step, freshwater becomes a little saltier on its way from the mountains, in the rivers, and to the sea. That's because on its long journey over and through the ground, it picks up salt from below the ground. In the rivers and lakes, and especially in the sea, some of the freshwater evaporates and salt is left behind. The salt sinks to the bottom of the oceans or ends up on the coast. It's tasty when added to food! Look around the next time you're in the supermarket to see if you can find sea salt on the shelves!

TRANSPORT
Clouds move from one place to another due to the force of the wind.
CONDENSATION
High in the air, the water vapor cools down again and makes little water droplets. These then form clouds.
TRANSPIRATION
Plants take water from the ground, then sweat it back out through their leaves as vapor.
EVAPORATION
INFILTRATION
Water falls to Earth as rain and penetrates the ground to become groundwater.
How long do water droplets stay in a cloud?
It normally takes about eight days before an evaporated water droplet falls back down to Earth as rain, hail, or snow.

WATER EVERYWHERE

You can drink water, you can cook with it, take a bath or shower in it, use it to flush the toilet, and to water the plants in your garden. Farmers need water for the animals to drink and to grow their grain. Factories also use lots of water: to cool machines, to make beer or fizzy drinks, or to wash trucks.

But where does all that water come from? We take some of the water from rivers and lakes. We call this surface water. Since it's easily accessible, it's used for lots of different things, from the transport of goods by ships to the cooling of power stations, from the irrigation of fields to making drinking water. But the total amount of surface water doesn't stay the same all the time. When it has been raining heavily for some time, the rivers are full of water. After a long dry period, there will be little water in the rivers. That's when we can't take much water out of the rivers, because that would threaten the animals that live here.

There is also lots of water in the ground, and we can pump up that groundwater. About 35 percent of the drinking water in the United States comes from groundwater because that is often the cleanest water source. You can see water in the landscape in many different ways worldwide!

Peat (the Netherlands)

Mine crater (Belgium)

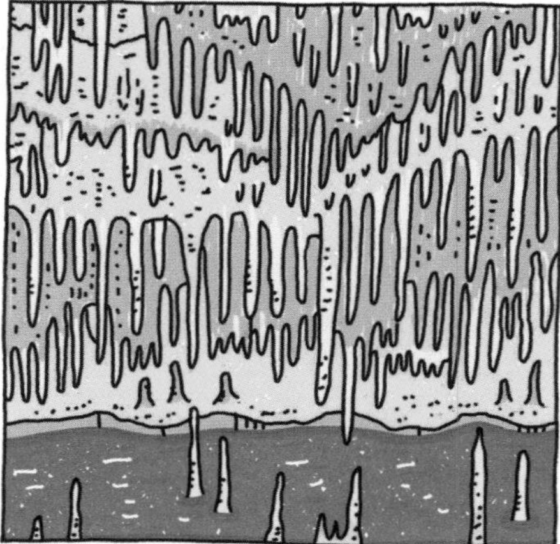

Devils Hole (Nevada, US)

Niagara Falls (USA/Canada)

Water tower (Belgium)

Vltava River (Czech Republic)

Jigokudani Snow Monkey Park (Japan)

Cenotes (Mexico)

Geyser (Iceland)

Boesmansgat (South Africa)

Pamukkale (Turkey)

Perito Moreno Glacier (Argentina)

Loch Ness (Scotland)

Polder (the Netherlands)

Can you drink groundwater right away?

Deep groundwater is often very clean. Almost no purification is needed to make that groundwater into drinking water. You find a lot more impurities in shallow groundwater. For example, muck can come from manure on the fields, chemicals used to keep weeds or rats out of the yard, or waste from factories and households.

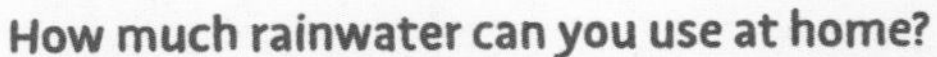

How much rainwater can you use at home?

If you have a water butt or tank, approximately half the water you use in your home and yard can be rainwater. That's good for the environment!

Whenever you need water to flush the toilet, wash your clothes or the dishes, clean the car, or water the plants in the yard, you can use rainwater. The water falls onto the roof of your house, then down a drainpipe into a water butt or a tank, usually under the ground, where you can store it until you need it. This helps you avoid wasting precious drinking water.

WHERE DOES GROUNDWATER COME FROM?

The main part of the water on our blue planet can be found in seas and glaciers. But you can't drink saltwater or ice. The water that's not salty or frozen is almost all underground. Groundwater doesn't flow in underground lakes or rivers. Instead, you find it in the little holes and cracks between grains of sand and between stones and rocks underground.

When it rains, part of the rain disappears into the ground. Those raindrops find their way into the groundwater. It can be just under our feet or much deeper, even down to hundreds of yards / meters deep. Some droplets reach the groundwater after a few days, weeks, or months, but it can also take years or even centuries.

Groundwater often remains in the ground for a very long time. But eventually all that water will flow into streams, rivers, and seas. There are different kinds of groundwater: shallow and deep. The shallow groundwater is located only a few yards / meters to a few dozen yards / meters under the surface. When it rains, groundwater is topped up. When it's dry for a while, the shallow groundwater goes down. The deep groundwater is located a hundred or even a few hundred yards / meters deep. It is very well protected there. For that groundwater, it doesn't matter much whether it rains a lot or very little.

It's very important that we don't pump up too much groundwater because, as you now know, it can take years or centuries before the groundwater is topped up with rain again. If we do use too much groundwater, rivers can

Is there any sparkling water under the ground?
The little bubbles in the sparkling water we drink are generally added to the water in a factory. However, there is also groundwater that contains natural bubbles. That water fizzes all by itself!

How long does water remain under the ground before it comes back up?
Once the droplets have reached the groundwater, their journey through the ground can still take days, weeks, months, years, or even thousands of years. If you were to follow the journey of a raindrop through the water cycle, you might grow very old in the process!

HOW DO YOU FIND GROUNDWATER?

Now you know what groundwater is and how groundwater reserves are topped up, but you don't know yet how to find that underground water. After all, you can't see the water deep beneath your feet. All the same, humans have been using it for thousands of years.

The landscape around us offers plenty of tips. You have more of a chance of finding groundwater at a modest depth in a valley than on top of a mountain. Some trees or plants that love water, such as willows or reeds, give away the places where groundwater can be found.

But how to you get it up? It's not easy! First you examine the geological map of the ground. That tells you what kinds of soil and stones are situated on top of or next to one another. In order to pump water up, you have to find stones with enough holes or tears for the water to gently flow through.

Found it? Then it's time to get down to work. Using a drill, you make a deep hole in the ground. A hundred yards / meters deep if necessary! And then…get pumping!

LET'S DO IT!

≈ Next time you go to the beach, take a spade with you.
≈ Look for a good place to dig a hole.
≈ If water appears in the hole after a while, then you've found groundwater!

Can your house sink?

When the ground is dry, the clay in the ground can shrink. If your house is built on a spot with shrinking clay, it can sink to one side or cracks can appear in the walls. Watch out!

Can you see groundwater from space?

The groundwater supply is like a bottle of water. A full bottle of water is heavy. If you drink until it's half empty, it gets lighter. The same is true of water on Earth. From space, the GRACE satellite can detect when the water supply under the ground becomes lighter.

How do you use electricity to find water in the ground?

Researchers sometimes send electricity into the ground with metal rods. Electricity moves more easily in water than in air, and you can use a detector to measure it. When there's water in the ground, you'll measure a stronger current. That's a handy trick!

THE ROLE OF WATER IN THE CLIMATE

Why does it rain more in some places than in others? Why is the climate by the sea different from that in the mountains or inland? In fact, what exactly is climate?

Well, climate is the weather measured at different moments over decades. The weather is made up of the temperature, wind, cloud cover, and rainfall or snowfall at a particular moment in time. The climate determines what clothes you buy in the stores each year. The weather determines what clothes you wear each day. So the climate is the weather we expect in a particular place in the world. There are a great many factors that determine the climate: the heat of the sun, warm and cold layers of air, the reflection of the sun's rays on land, and so on.

The water on Earth is important not only because it brings rain but also because it stores the heat of the sun. Water warms up more slowly than land. That's why it's cooler by a swimming pool than on the street when the sun is shining. Therefore, in the summer, it's always a little cooler at the seaside than farther away from the coast. In the winter, it's the other way around. Then the sea slowly gives off its warmth, so it stays a little warmer at the seaside.

The huge quantity of water in the oceans flows from one side of Earth to the other. If you were a fish, you'd be able to go along for the ride. Those currents also shift heat from warm regions to colder places. Countries that are located close to these currents are always a few degrees warmer than countries that lie much farther away.

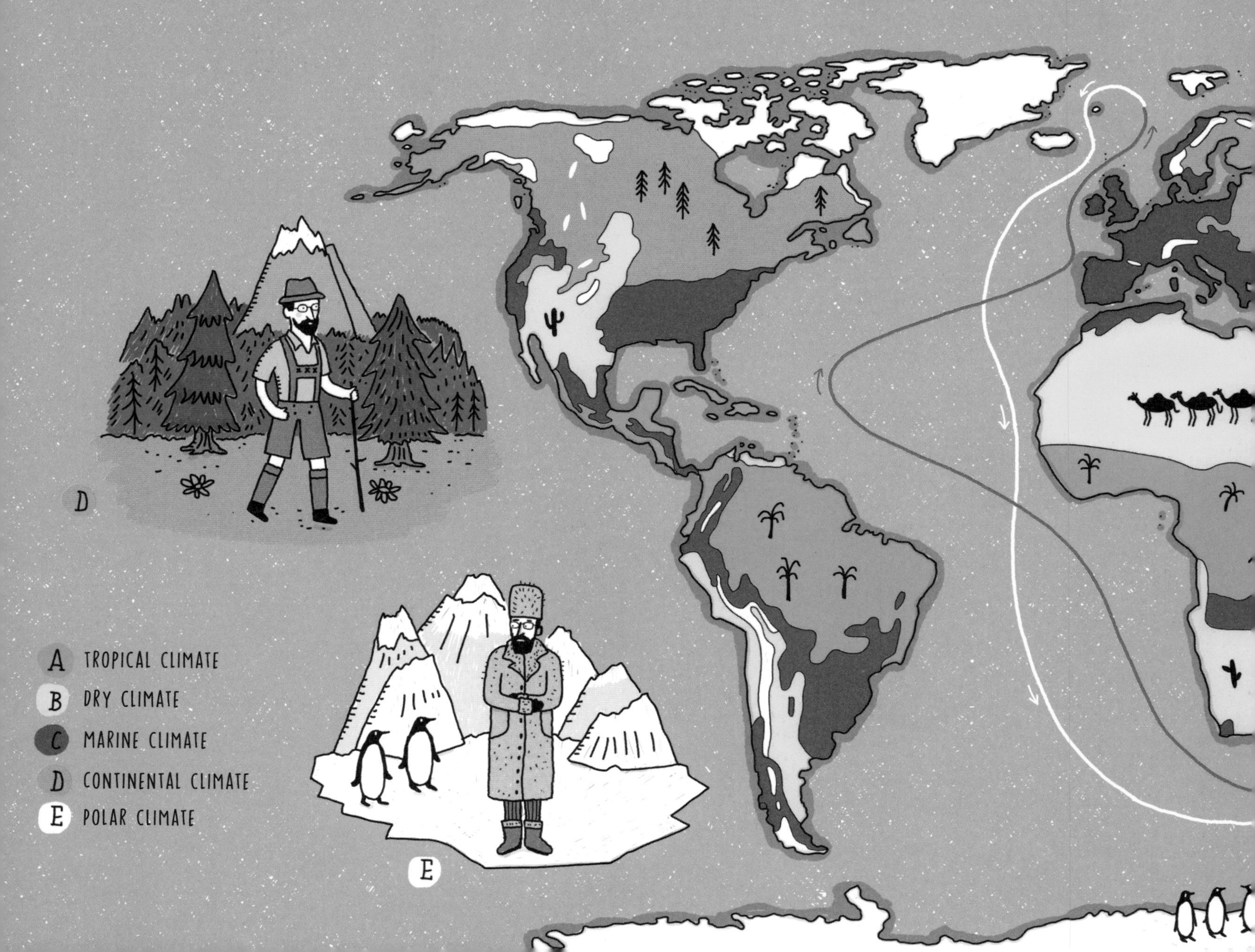

Mountains also play a role in the climate in any given place. On high mountains, there is often one side where it rains a lot and one where it rains very little, if at all. The air above us goes up several miles / kilometers and consists of different layers that grow colder as you go higher up. The wet air blown along by the wind has to move from low, warm layers of air to high, cold layers because the mountain is blocking the way. In the colder air, the water vapor turns back into water. The droplets form clouds, and eventually it begins to rain. Therefore, a chain of mountains often has a wet side (windward side) and a dry side (lee side). The fact that there is less rainfall on the lee side is called the rain shadow.

Who was Wladimir Köppen and what does he have to do with the climate?

We needed a system to describe the different types of climate on Earth. Wladimir Köppen came up with five major climate groups. He can't have been feeling very creative that day, because he simply named them with the letters A, B, C, D, and E. He used the temperature, rainfall, and plant species that can grow in a region as indicators to place the local climate in one of the groups. We still use his system today.

WATER AND THE WEATHER

Is it raining outside? Where are my rubber boots and my umbrella? Most of us are familiar with rain, but water can fall from the sky in various ways. When raindrops freeze, we get bombarded by hail. In the winter, the water sometimes falls as ice crystals. That's snow. It's so much fun to press that wet snow together into a beautiful snowball!

In the mornings in some places, you can see a different form of water: mist. Mist is made up of clouds floating just above the ground. Sometimes you can barely see your feet. A cloud is a collection of tiny water droplets, ice crystals, or a mixture of the two. Did you know that there are lots of different kinds of clouds and that they are constantly changing shape due to air currents?

But when does it actually start to rain? After all, there's no giant shower in space for someone to turn on. So what actually happens? The clouds floating above us in the sky are a collection of billions of tiny water droplets. In warm countries, it begins to rain because those tiny droplets in the clouds collide with one another, then clump together until they form a bigger droplet that is heavy enough to fall.

In colder regions, it gets so cold high in the air that you would even find ice crystals in the clouds. Those ice crystals grow bigger and bigger, and the droplets in the cloud freeze onto the ice crystals and stick to them. When the crystals grow heavy enough, they fall. First they fall as snowflakes, but because they generally melt before they reach the ground, we tend to see them arrive as rain.

Can you drink rainwater?

Rainwater looks perfectly clean, but you can't just drink it. Due to car fumes and factoy output, the air contains a great many substances that are bad for your health. The raindrops absorb some of those substances. There can also be a lot of dirt on the roof of your house. The water takes that dirt with it when it flows into your water butt or tank. So it's not a good idea to drink rainwater as it is. You never know, you might be drinking pigeon droppings!

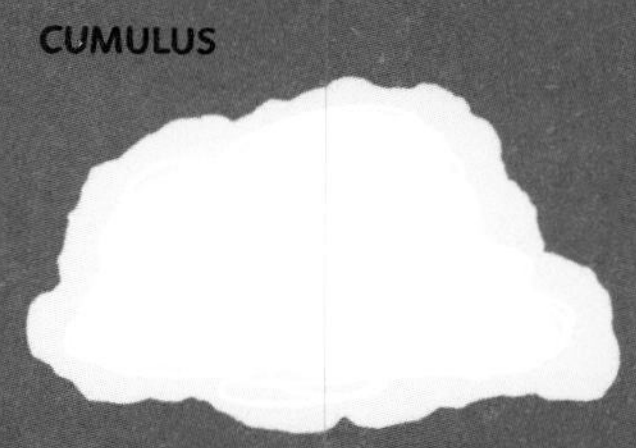

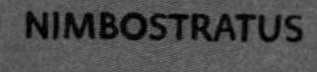

What is a cloud family?

In the 18th century, chemist Luke Howard often lay on his back in the grass looking up at the clouds and he realized that there were different types of clouds. Clouds of the same family are always found at the same height in the sky. Luke grouped clouds into high, medium, and low clouds, as well as towering clouds. He gave the different types of clouds funny-sounding names such as cirrus, stratus, and cumulonimbus.

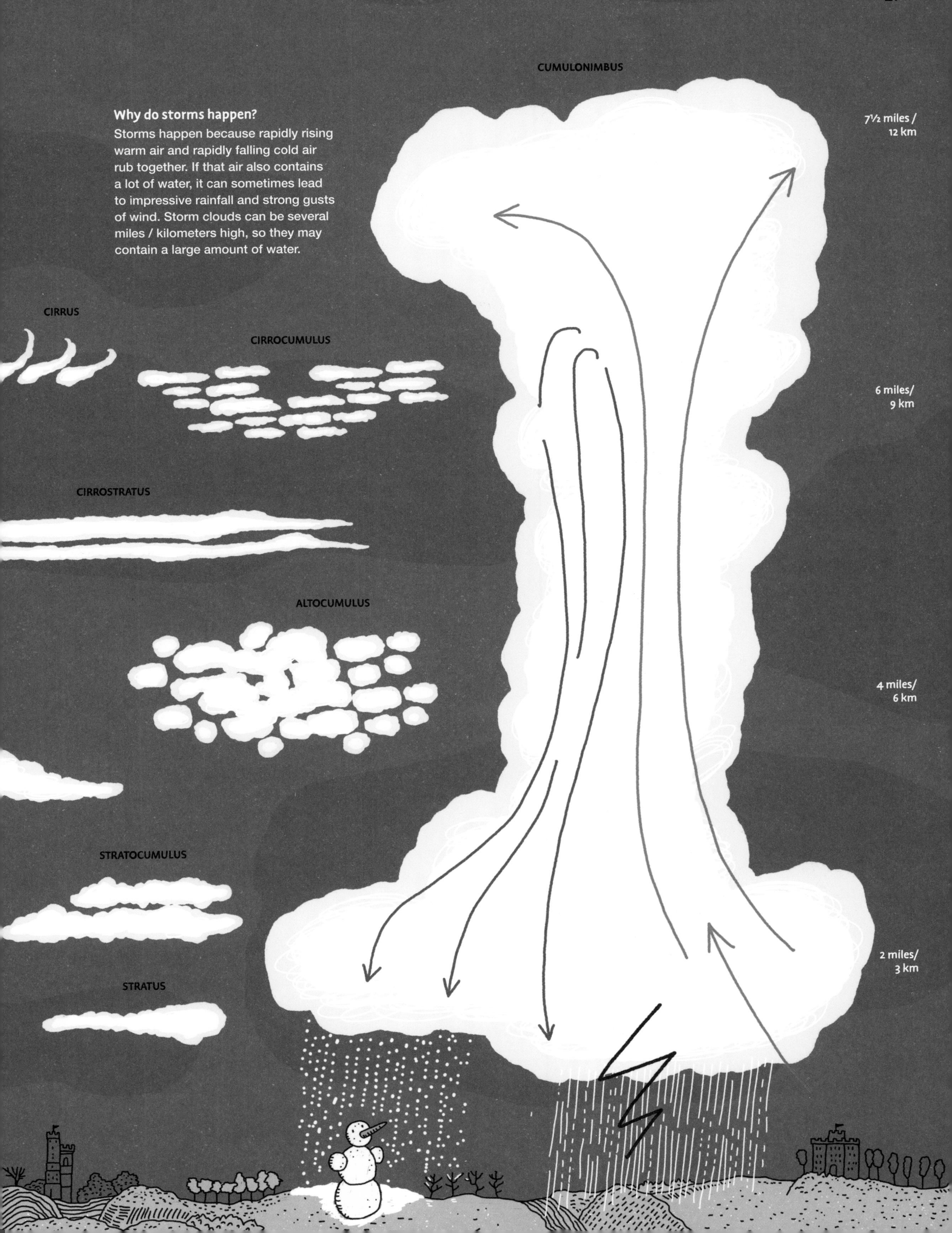

Why do storms happen?

Storms happen because rapidly rising warm air and rapidly falling cold air rub together. If that air also contains a lot of water, it can sometimes lead to impressive rainfall and strong gusts of wind. Storm clouds can be several miles / kilometers high, so they may contain a large amount of water.

HOW DID PEOPLE DEAL WITH WATER IN THE PAST?

Water is very important for us humans. You can understand that when you look back in time too. Major populations and civilizations all over the world invented clever ways of getting water to the places where they lived and farmed. Now we can sometimes find out precisely how they did that. Many of the inventions and devices we take for granted today were actually invented a very long time ago.

Even before 3000 BCE, people in Mesopotamia in the Middle East dug canals to bring water to their fields. It was very dry there, but because the rivers Tigris and Euphrates sometimes flooded, leaving behind fine mud or silt, the ground was actually quite good for growing plants. People made canals that they could open and close. They also used a shadoof—a lever with a bucket—to help them move the water from the river into the canals and to their land.

The Romans are known for their magnificent baths and their grand network of aqueducts. They used these large stone structures to bring clean water from the river into the cities. This helped prevent people from getting sick. The Romans weren't the ones who invented aqueducts, but they were masters in building them. You can admire one in the south of France, more than 2,000 years after it was first built: the Pont du Gard.

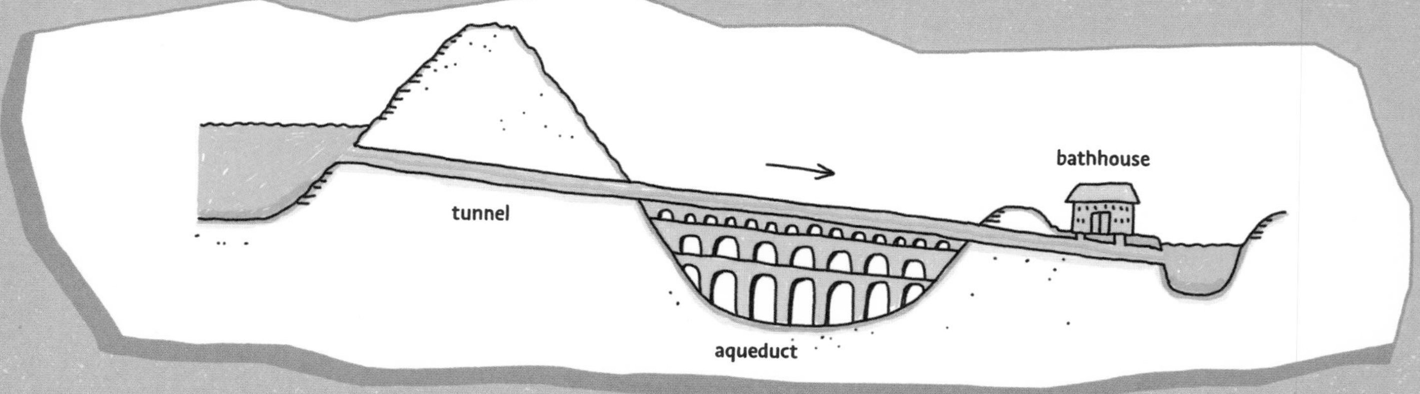

The Gardens of Nineveh

Archimedes had his best ideas in the bath. Eureka!

The Ancient Greeks wrote about a city with terraces filled with magnificent plants. Was it Babylon or was it Nineveh? The locals took water from the mountains using underground canals and aqueducts. They used a water screw to transport it to the high gardens. The Greek scientist Archimedes is said to have invented the water screw, but it probably existed before his time.

water clock

Can you make a clock out of water?
In Ancient Egypt, as well as in other countries such as Persia and China, people didn't just use water to grow food, they also made clocks that ran on water. Using a system of pipes and reservoirs, they were able to measure how many hours went by. This was useful, for example, in the courts to ensure that the lawyers didn't talk for longer than was agreed, or for priests who wanted to pray to the gods at the right time of day.

Harappa sewer

How did people get water to high places in the past?
Anyone wanting to get water from a low place to a high place needs a pump. At least 3,000 years ago in Ancient China and Egypt, people hung buckets onto large wheels to make scoop wheels. These were the first type of pumps. By turning the wheels, they could quickly scoop water out of the river and transport it to a canal higher up. People and animals had to use their strength to turn the wheels.

When were the first sewers built?
The people of the Indus Valley civilization had a whole sewage system below their cities more than 5,000 (!) years ago, neatly built with bricks.

WHY DO WE NEED TO DRINK WATER?

An adult needs around 6 to 8 cups / 1.5 to 2 liters of water a day. That's one large bottle. Precisely how much you need depends on how healthy and how old you are, how hot it is, what you're doing, and what you eat. If you do sports, you can lose as much as 4 to 8 cups / 1 to 2 liters of moisture in an hour! Your body needs the water for various purposes: to transport nutrients, to protect it against shocks, and to regulate your temperature. When you sweat your skin becomes wet. Those sweat droplets evaporate, cooling your skin.

More than half of the human body is made up of water! We lose the most water by going to the toilet, but we also lose water by sweating and breathing. On average, a healthy adult loses around 11 cups / 2.5 liters of liquid a day. You take in around 4 cups / 1 liter of liquid from your food each day. The rest of the water we lose has to be topped off by drinking. If you don't drink enough, your body will soon let you know that something is wrong. At first you get thirsty, then you start feeling groggy and weak, and eventually you become very confused and suffer from headaches.

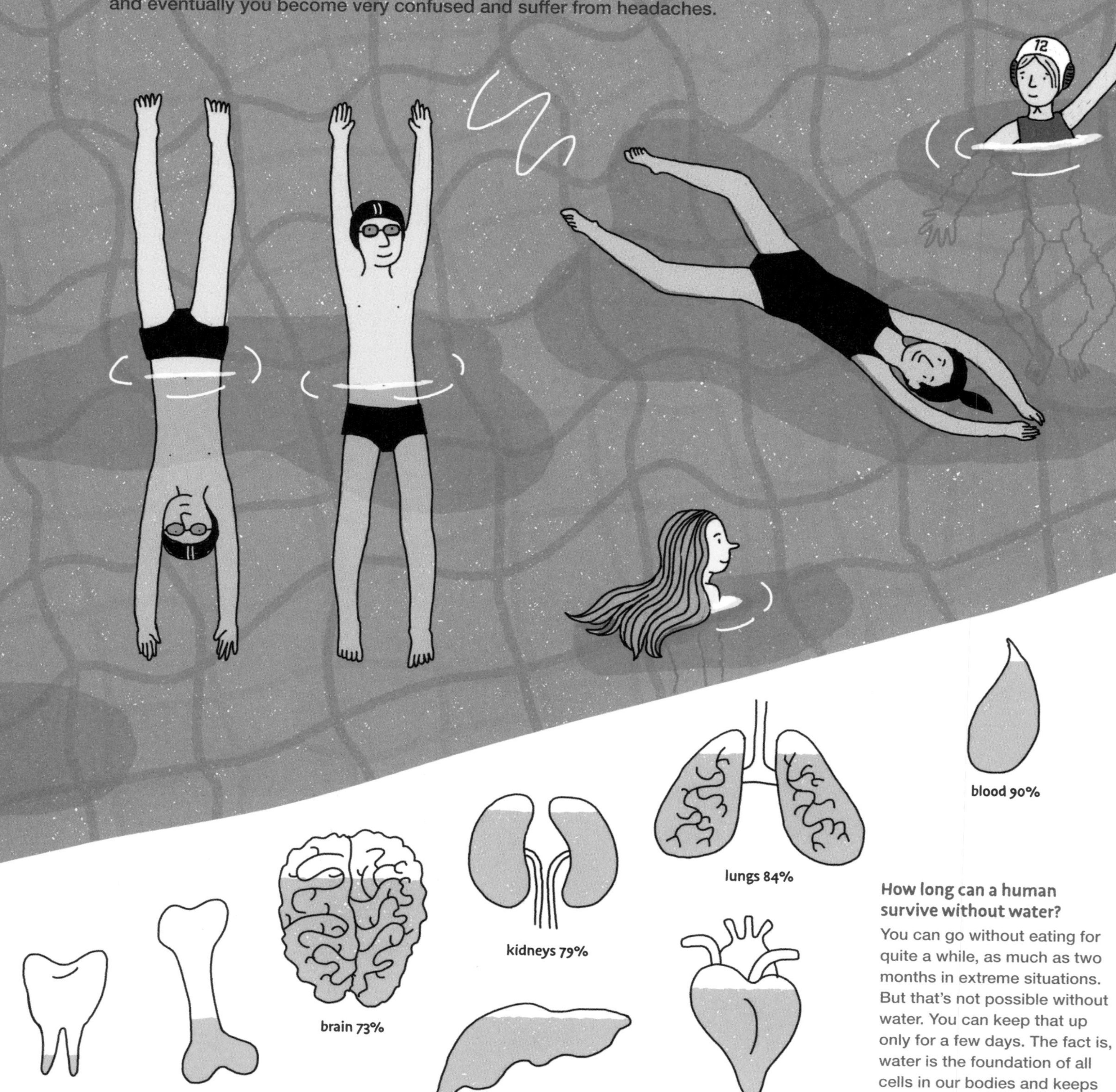

How long can a human survive without water?

You can go without eating for quite a while, as much as two months in extreme situations. But that's not possible without water. You can keep that up only for a few days. The fact is, water is the foundation of all cells in our bodies and keeps our organs working.

Why do some people float better in water than others?

Some people float easily on water, and others sink all the time. It's got nothing to do with your size, and everything to do with the density of your body. The density is a measure of how closely the particles that make up an object are stuck together. Wood can float on water, but a stone cannot, because the stone has a higher density. Some people are more like wood, and others more like a stone. One thing's for sure: Fat floats better than muscle. You also float better if you fill your lungs with air first.

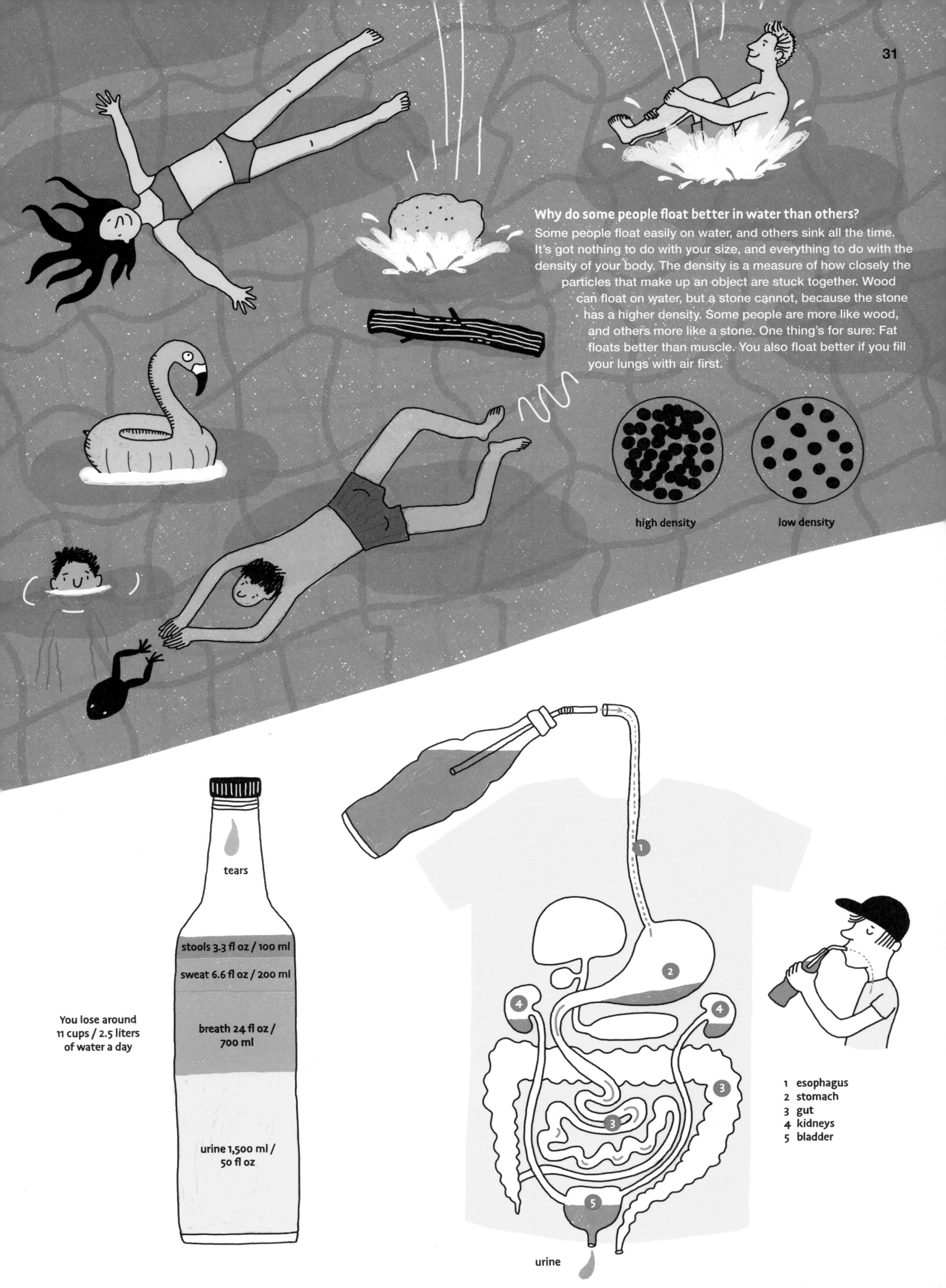

HOW DOES WATER PURIFICATION ACTUALLY WORK?

We use all kinds of filters and other clever ideas to clean water that we take from our natural environment. First, the water purification plant takes the large pieces of dirt, such as dead fish or pieces of wood, from the water with a coarse filter. It works just like a fishing net that you use on the beach to catch crabs or shrimp.

The filtered water then goes into a reservoir, where the larger pieces of dirt sink to the bottom. Because the water there is still, heavier pieces of dirt sink slowly. In the reservoir, nature gets down to business. The small animals swimming around in the water use the dirt for food, so nature does the job for us.

The water is then pumped up and pushed through very fine filters. To deal with the dirt that is filtered out during this process, the treatment plant uses a special chemical to get the dirt particles to clump together. The lumps are pushed upward with something that looks like a giant underwater bubble blower so that they will float. The floating scum can now easily be skimmed off. Next, the water goes through a sand filter, where the final tiny dirt particles are caught between the grains of sand.

Now the only dirt remaining in the water is dirt that you can't see. The water purification plant has found a new superweapon to deal with that: an activated charcoal filter, which strongly attracts invisible flavor and fragrance chemicals, medicines, and other waste materials, holding on to them like a sponge. When the water comes out of this last filter, it's completely pure.

Finally, the water is disinfected with chlorine. Chlorine is a miracle remedy for dealing with bacteria. Chlorine is also used in most swimming pools to make sure you don't get sick from the swimming pool water. At last, the disinfected water is ready to drink.

Can dunes filter water?

A sand filter doesn't have to be artificial. Some water purification plants use the sand of the dunes to filter water. The water runs through the dunes and is cleaned because the pieces of dirt get stuck in the tiny spaces between the grains of dune sand.

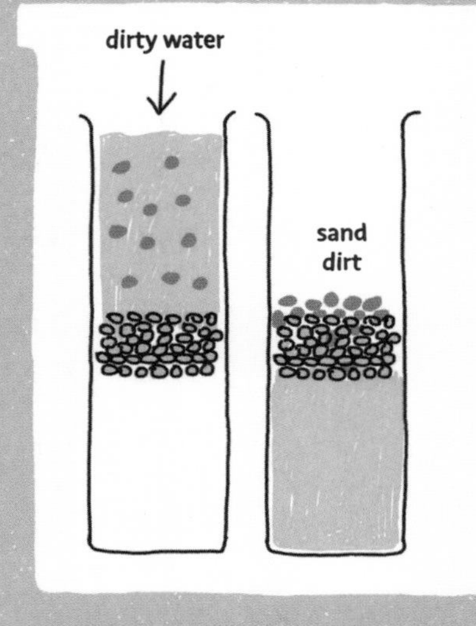

What is activated charcoal?

Activated charcoal is not for eating. It looks a bit like the charcoal you use on a barbecue. The black substance has thousands of tiny holes that you can't see, but they can catch the dirt particles.

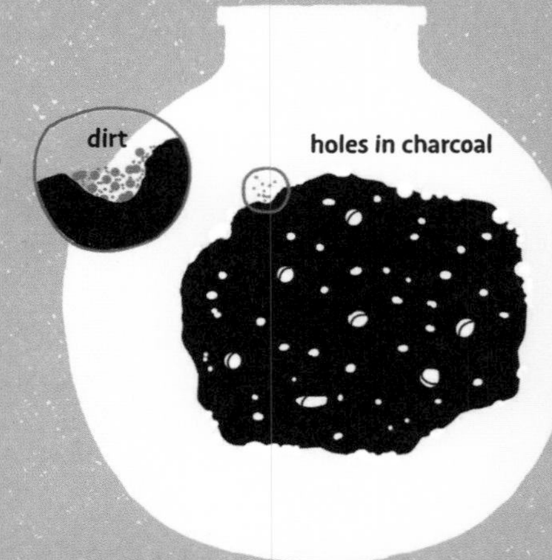

LET'S DO IT!

≈ Add some sand or mud to water and stir hard.
≈ Let everything stand for about half an hour.
≈ What do you see when you look at it? It's just like the water purification reservoir. The pieces of dirt have sunk to the bottom.

James Simpson
inventor of the sand filter

What did people drink in medieval times?

They drank beer! The brewer would boil the mixture of water, grains, and hops to make beer, and that also killed the germs. In those days, water was often infected with diseases because people did not yet have good sewage or water purification systems.

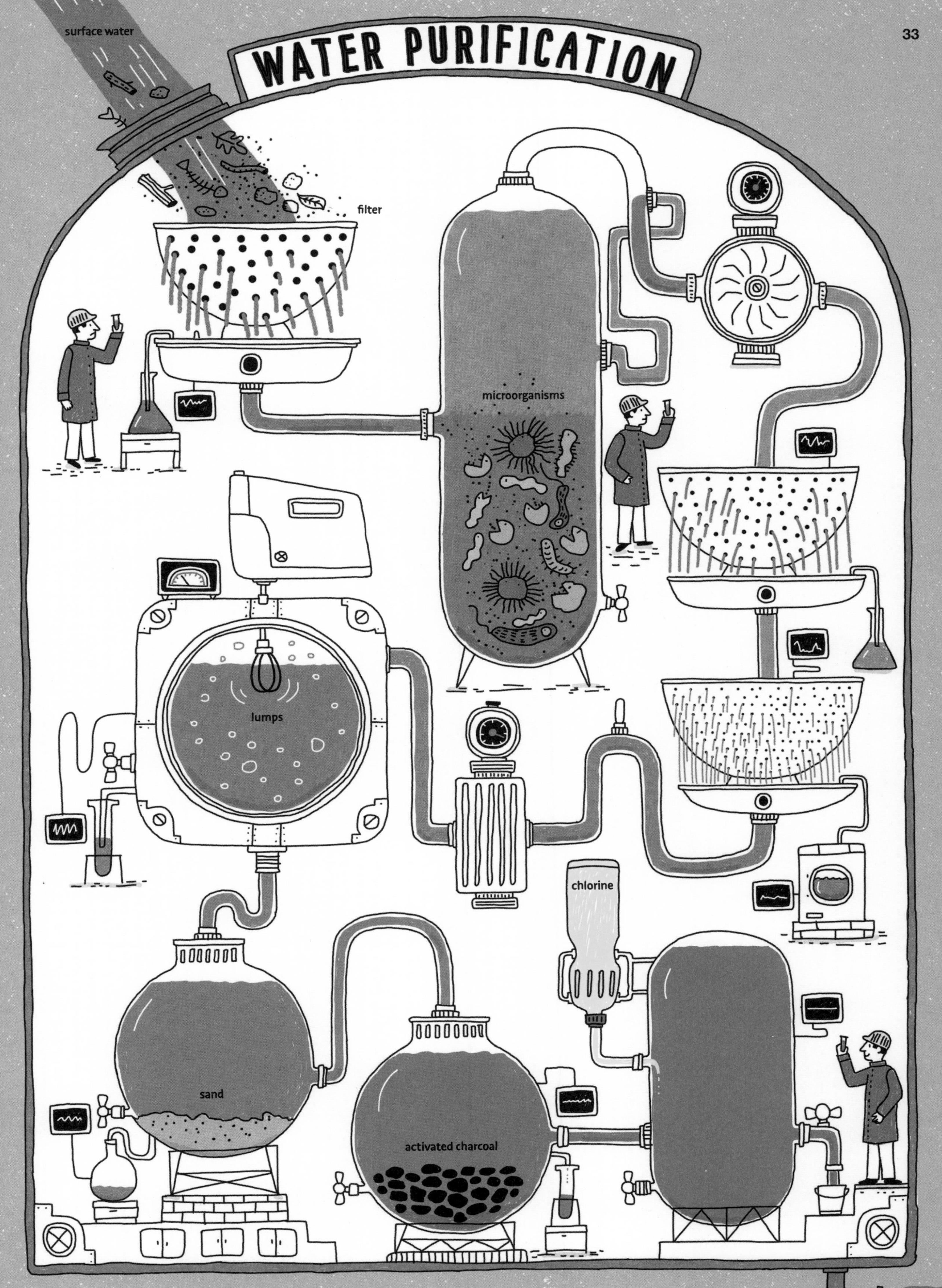
surface water
WATER PURIFICATION
filter
microorganisms
lumps
chlorine
sand
activated charcoal
drinking water

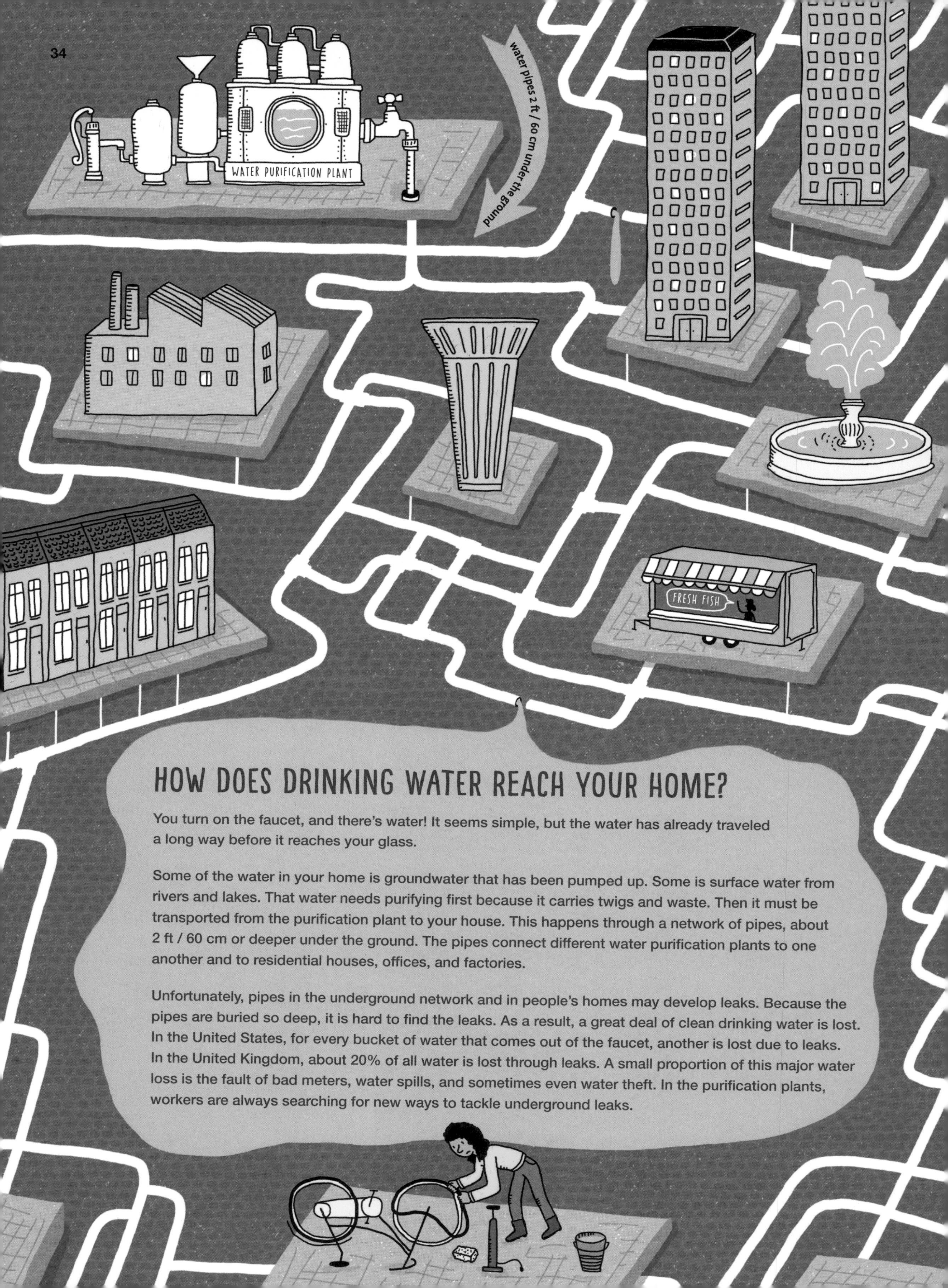

HOW DOES DRINKING WATER REACH YOUR HOME?

You turn on the faucet, and there's water! It seems simple, but the water has already traveled a long way before it reaches your glass.

Some of the water in your home is groundwater that has been pumped up. Some is surface water from rivers and lakes. That water needs purifying first because it carries twigs and waste. Then it must be transported from the purification plant to your house. This happens through a network of pipes, about 2 ft / 60 cm or deeper under the ground. The pipes connect different water purification plants to one another and to residential houses, offices, and factories.

Unfortunately, pipes in the underground network and in people's homes may develop leaks. Because the pipes are buried so deep, it is hard to find the leaks. As a result, a great deal of clean drinking water is lost. In the United States, for every bucket of water that comes out of the faucet, another is lost due to leaks. In the United Kingdom, about 20% of all water is lost through leaks. A small proportion of this major water loss is the fault of bad meters, water spills, and sometimes even water theft. In the purification plants, workers are always searching for new ways to tackle underground leaks.

What is a sinkhole?

If underground leaks go unnoticed for a long time, you can end up with spectacular sinkholes. Under a street or a building, the water from a leaking pipe washes away more and more ground. You might not see anything for quite a while, because the road remains in place, like a bridge over a deep valley. But suddenly the hole becomes too big, and the road or building can't support itself any longer. Then it collapses into the hole. A sinkhole can be so big that entire cars or houses are dragged down.

Why does water from the faucet contain chalk in some places?

During its journey through the ground, groundwater absorbs particles from the soil such as chalk. Chalk is healthy, but it can leave white marks in your shower or break washing machines. If the tap water in your house comes from groundwater, there can be a lot of chalk in it.

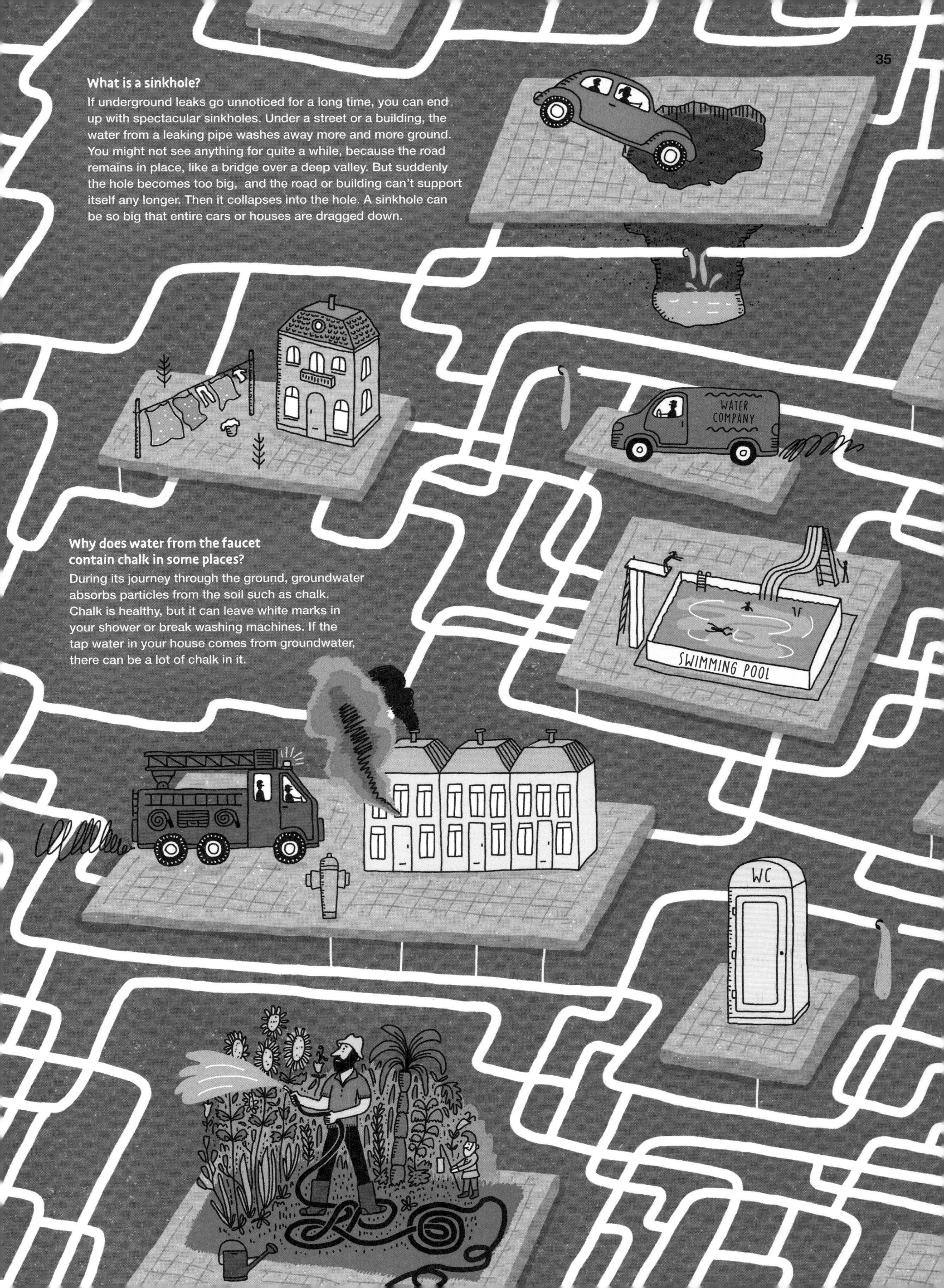

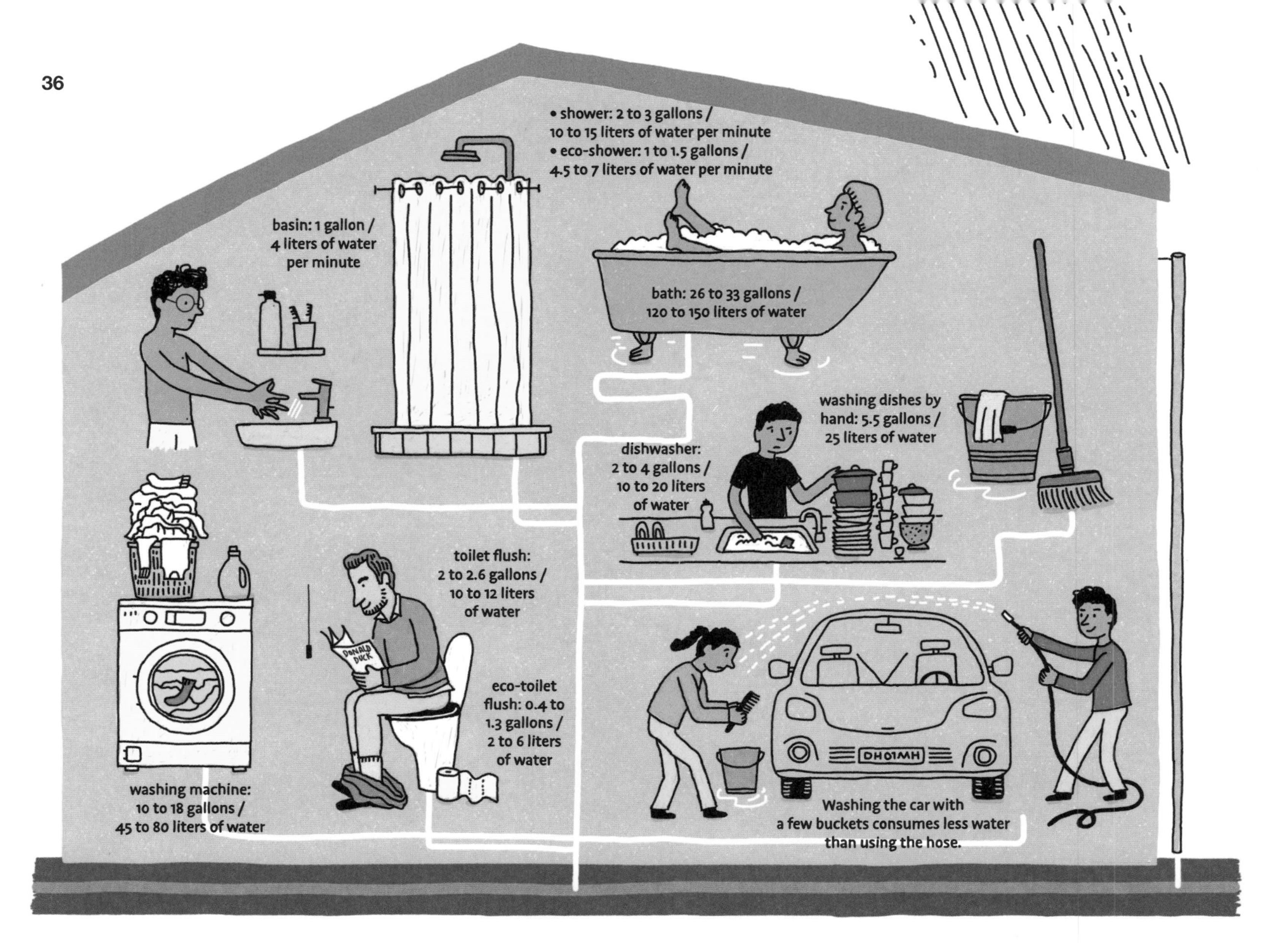

WHAT HAPPENS TO YOUR URINE?

You've been to the bathroom, and you flush the toilet. Your urine is about to go on an extraordinary adventure. Just like at a water park, the water swooshes down through the pipes in your house and splashes out into the sewer. There it begins a great long journey through bends, collection chambers, and busy underground intersections via even bigger pipes to the wastewater treatment plant.

Like in the drinking water purification plant, the wastewater treatment plant has various filters for collecting dirt. After that, the water goes into large, round basins where tiny creatures, or microorganisms, have the time to eat up the dirt in the water. Some of these creatures are already in the water, but most are in the mud layer at the bottom of the water basins. Finally, the pieces of dirt in the water sink to the bottom and form a layer of mud. Only after that has happened, does the water go into the river to start the whole process all over again.

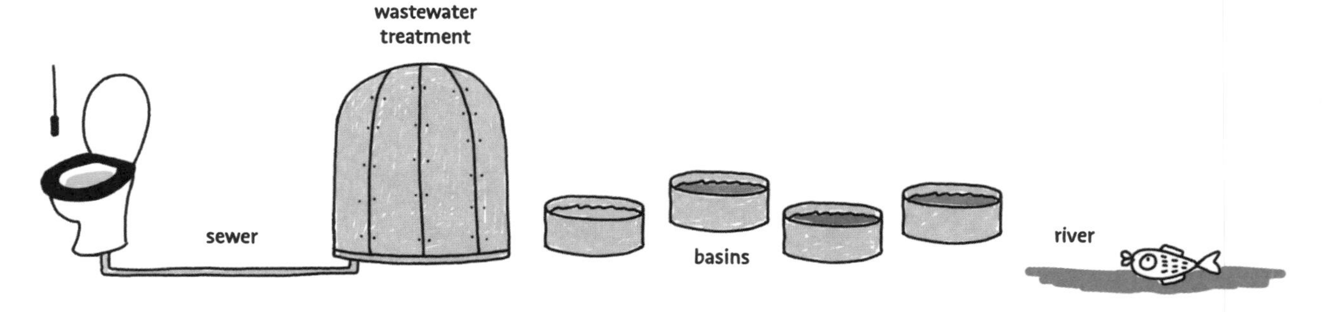

In the past, a great deal of the wastewater from the sewers went straight into streams and rivers. As you can imagine, it smelled awful! Fish and other aquatic animals went in search of cleaner places to live. In many countries, people still flush the toilet water straight into the river, which can make other people sick if they use the water farther downstream. That's why lots of countries are working hard to collect the used water from the houses and treat it before it goes back into the rivers.

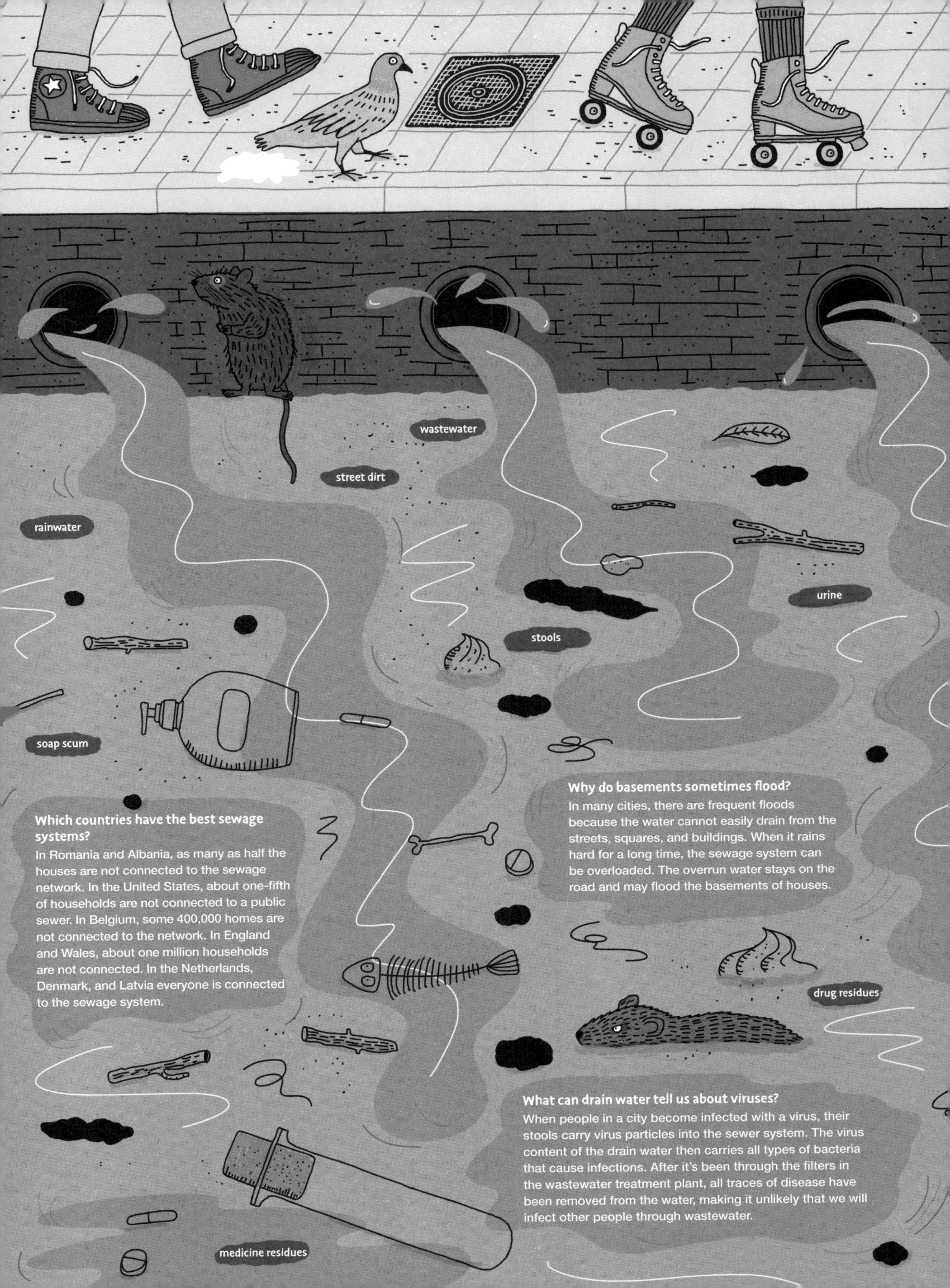

Which countries have the best sewage systems?

In Romania and Albania, as many as half the houses are not connected to the sewage network. In the United States, about one-fifth of households are not connected to a public sewer. In Belgium, some 400,000 homes are not connected to the network. In England and Wales, about one million households are not connected. In the Netherlands, Denmark, and Latvia everyone is connected to the sewage system.

Why do basements sometimes flood?

In many cities, there are frequent floods because the water cannot easily drain from the streets, squares, and buildings. When it rains hard for a long time, the sewage system can be overloaded. The overrun water stays on the road and may flood the basements of houses.

What can drain water tell us about viruses?

When people in a city become infected with a virus, their stools carry virus particles into the sewer system. The virus content of the drain water then carries all types of bacteria that cause infections. After it's been through the filters in the wastewater treatment plant, all traces of disease have been removed from the water, making it unlikely that we will infect other people through wastewater.

WITHOUT WATER YOUR PLATE WOULD BE EMPTY

Farmers need a great deal of water to provide food for us. In many regions, such as the Midwest in the United States or Belgium and the Netherlands in Europe, people can count on plenty of rain. Still, climate change means that there is less rainfall and that it is less well distributed throughout the year.

When the land is dry, farmers have to water the fields themselves. We call that irrigation, and it can be done in various ways—for instance by flooding the fields or using sprinklers and irrigation tubes. This water obviously has to come from somewhere else, such as the ground or a river, or perhaps from a storage tank where farmers can store water from heavy winter rainfall.

Too much water, on the other hand, can also be problematic for farmers. Most plants don't like having their roots saturated with water. That's why a lot of land needs to be drained. It means that surplus floodwater in the ground is transported away via streams, canals, or even pipes with holes underneath the fields. If we didn't remove the surplus water, it would not be possible to use some areas of land for farming. That land would simply turn back into marsh and wetland. Much of the land was reclaimed long ago by using large waterworks in order to produce enough food.

It's a shame that so much valuable water is transported to the rivers where it immediately flows out to sea, rather than being stored for later. We should think carefully about where and when we allow the water to flow away so that we waste as little as possible of the water that we will certainly need in dry periods or times of serious drought.

Which farms need the most water?

No two farmers or farms are the same, and they don't all need the same amount of water. Vegetables and fruit grown in greenhouses require the most water, but a great deal of water is also needed for cultivating certain crops in the fields and for the production of beef and pork. You can read more about that on the next page.

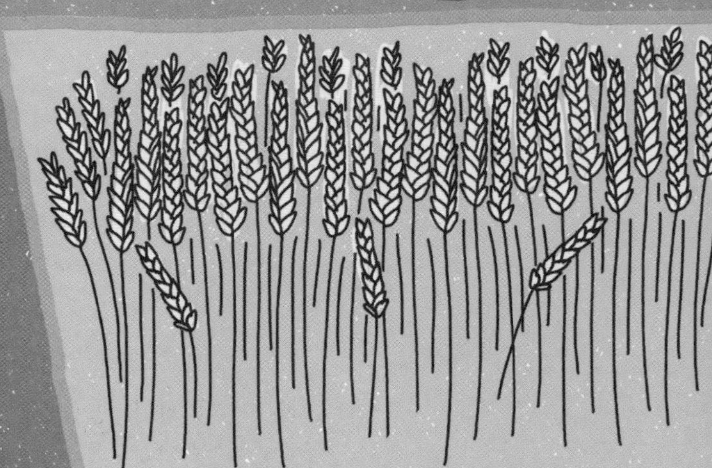

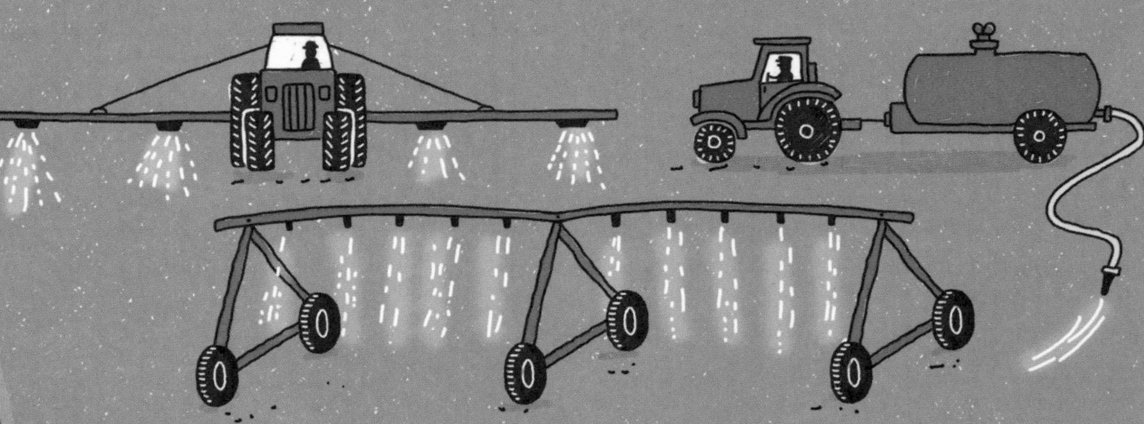

Does farming always use lots of water?

In Western Europe, there is generally enough rain so that farming doesn't need a lot of extra water. It's a different situation in countries where it doesn't rain during the growing season. If there is suddenly a water shortage, there can be major consequences. When the weather stays dry for too long, plants or animals can die and months of the farmers' work can be lost.

WHAT IS VIRTUAL WATER?

We call all the water needed to make a product virtual water. Water is needed for almost everything we use or eat. Think of a steak, for example. That hunk of meat comes from a cow's sturdy rump. She needs to drink throughout her life, and all the more so if she also produces milk. A cow needs to drink at least ten buckets of water a day, and the cowshed has to be cleaned.

You know that a cow also has to eat. Her tasty hay or other feed has to grow somewhere, and for these plants to grow, water is also needed. So you can imagine that a great deal of water goes into producing that steak. Of course, it depends how the farmer goes about it, but at least 440 to 660 gallons / 2,000 to 3,000 liters of water (or around 200 to 300 buckets) are needed for a steak of about 7 ounces / 200 grams.

The water doesn't always come from a stream beside the field. Not all farmers grow the feed for their animals on their own land, but a lot of water is needed to grow the grass, clover or soy for our cow somewhere in the world.

In fact, water is not only required for producing our food, it is needed to make pretty much everything, from a T-shirt to a phone. All the products we use have a virtual "water footprint." That water footprint is the sum total of all the water needed to make an item: It's not only about the water that is consumed, because the water that is polluted in making a product is also part of the water footprint.

Who invented the water footprint?

The Dutch professor Arjen Hoekstra invented the idea of a water footprint. Many products are manufactured in a place other than where they are sold. Take a look at the label in your T-shirt. Where was it made? By counting up all the water needed to make a product, even when that water was used somewhere else in the world, Arjen Hoekstra showed us that we use a great deal more virtual water than we think we do. Just look at your water bill.

1 apple: 27.5 gallons / 125 litres

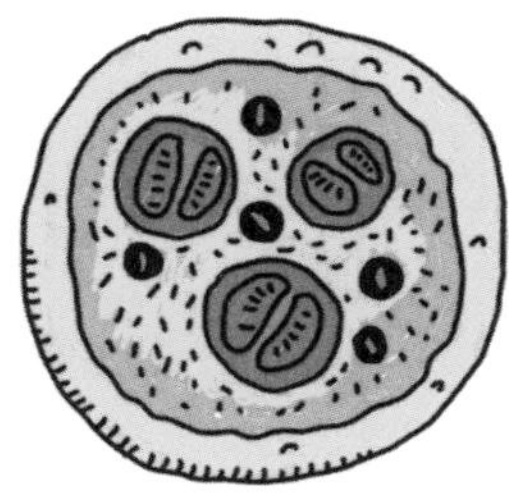

1 pizza: 277 gallons / 1,260 liters

1 tomato: 11 gallons / 50 liters

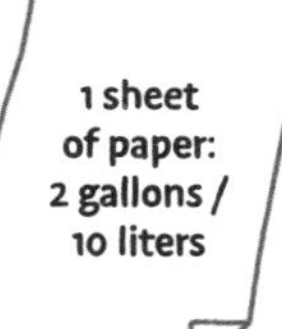

1 sheet of paper: 2 gallons / 10 liters

Arjen Hoekstra engineer

1 pair of jeans: 2,387 gallons / 10,850 liters

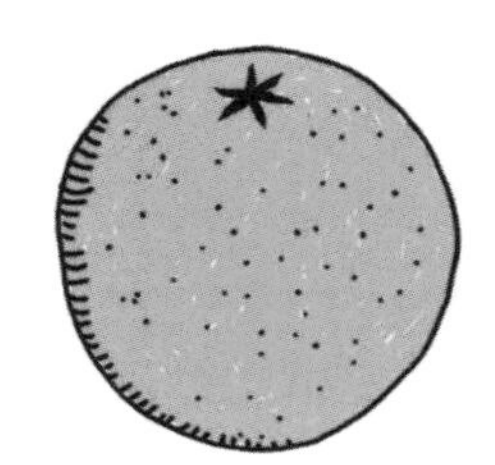

1 orange: 17.6 gallons / 80 liters

1 smartphone: 200 gallons / 910 liters

1 cup of tea: 6 gallons / 27 liters

1 chocolate bar: 880 gallons / 4,000 liters

2.2 lb. / 1 kg pork: 1,318 gallons / 5,990 liters

1 T-shirt: 598 gallons / 2,720 liters

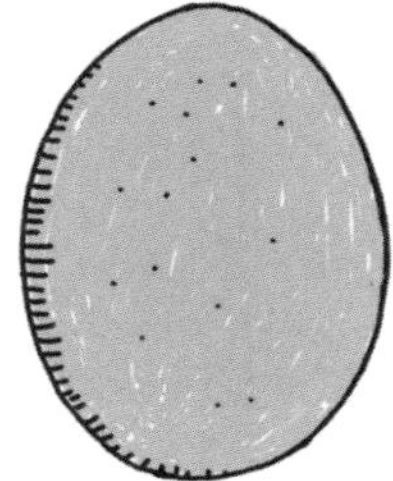

1 egg: 43 gallons / 200 liters

2.2 lb. / 1 kg poultry: 951 gallons / 4,325 liters

2.2 lb. / 1 kg cheese: 699 gallons / 3,178 liters

What is blue, green, and gray water?

The water footprint includes three kinds of water: green water (rainwater), blue water (rivers and groundwater), and gray water (water that has previously been used for something else). The water doesn't really have those colors, but it's a code people use to talk about the different types of water sources.

LET'S DO IT!

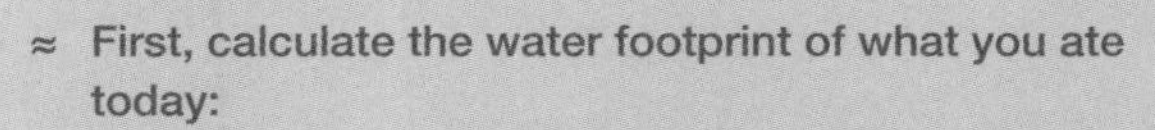

≈ First, calculate the water footprint of what you ate today: https://waterfootprint.org/en/resources/interactive-tools/product-gallery

≈ Now calculate your own water footprint: https://www.watercalculator.org/

≈ Do you use more than is sustainable and easily available to everyone? Then you should find out how to use less virtual water.

HOW DO PLANTS TAKE WATER OUT OF THE GROUND?

Tomatoes need lots of water, cactuses need less, but no plant can do without water altogether. The ground is like a wet towel that has been squeezed dry: It still feels damp, but the water doesn't just trickle out of it. So there's still lots of water in it, but we can't get it out easily. A plant can, though! That's why they need their roots. The roots can sometimes search for water several yards / meters underground. The roots also prevent the plant from falling over by anchoring it firmly in the ground.

The roots and stem of a plant can be seen as a collection of tiny straws. One end of the straw is in the ground where the water is stored. At the other end of the straw are the leaves, which suck on the straws to get at the water.

The leaves ensure that the plant doesn't get too hot. You sweat when you get too hot, and that makes you thirsty. It's the same for plants. Lots of water evaporates from their leaves. We call that transpiration. It has the same effect as sucking on a straw: The water is sucked up out of the ground through the roots and the stem. On the way to the leaves, water also brings nutrients to the plant, like boats on a river.

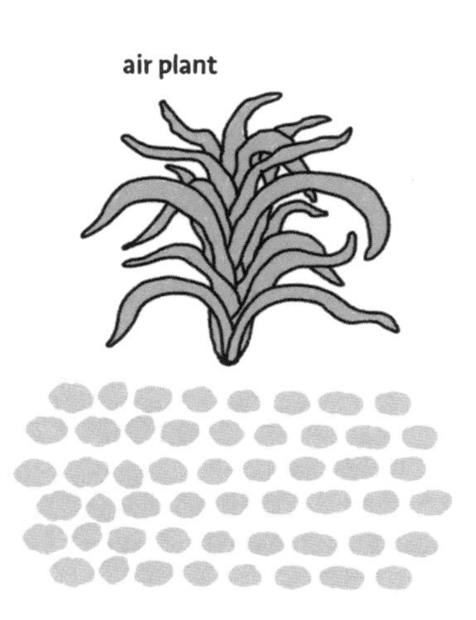

Are there any plants without roots?

Some plants can take water and nutrients from the air, rain, and dust through the scales on their leaves. These air plants grow very slowly. Most air plants in garden centers come from the southern United States and Central America. They are taken from their natural environment to become houseplants. This has made some species very rare in nature.

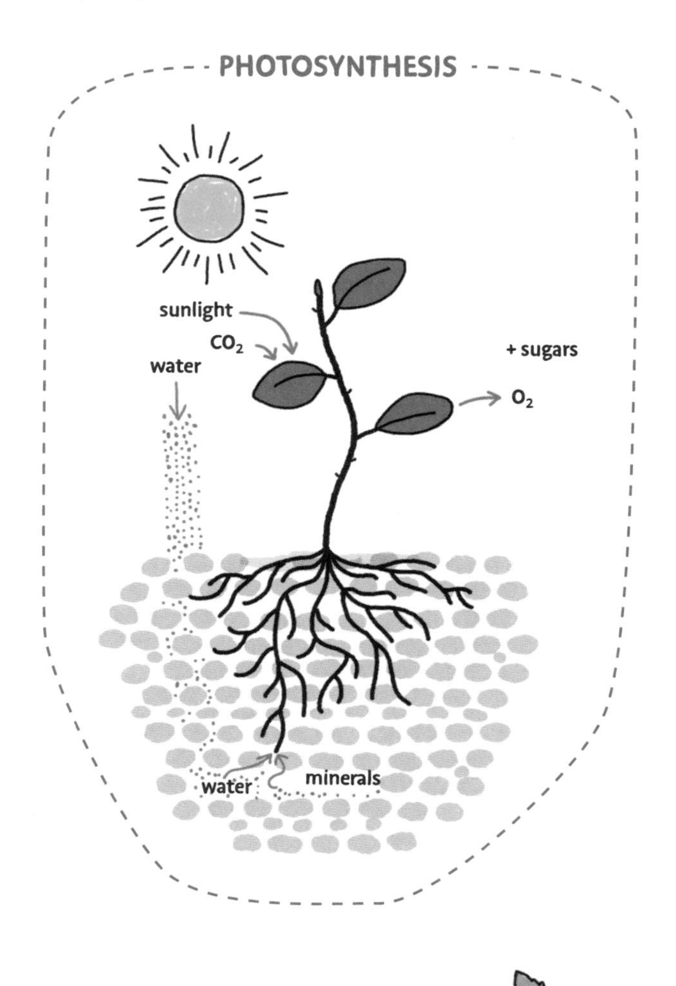

Why is a tree trunk thicker during the day?

Plants take up water during the day but not at night, because they need light to convert the water and carbon dioxide into food. That's why the trunk of a tree becomes thicker during the day and thinner again at night. By measuring the thickness of the trunk, we can learn a bit about how much water a tree takes up and when it does so. If the trunk doesn't grow thick enough during the day, we know that there is probably not enough water in the ground.

LET'S DO IT!

Cut a twig from a tree, wait for a few minutes, then see what appears on the cut surface. When you cut off a twig, the leaves can no longer properly suck up water into the branch, and it comes out of the cut surface as little droplets.

What are leaf pores?

There are lots of little openings on a leaf that the plant can open and shut. We call these leaf pores, or stomata. When it's very dry and there's not enough water in the ground, the plant closes its pores so that no more water will be lost through the leaves. When that happens, the plant feels unwell.

leaf pores

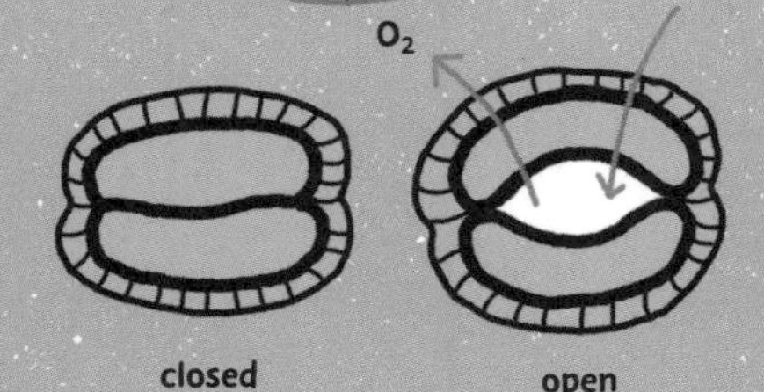

vessels

water and minerals on their way to the leaf

What does mold have to do with trees?

Plants work closely together with fungi and mold in the ground to find water and nutrients more easily. They can help one another get to the food and water they need, even when it's dry.

root

sucks water out of the ground

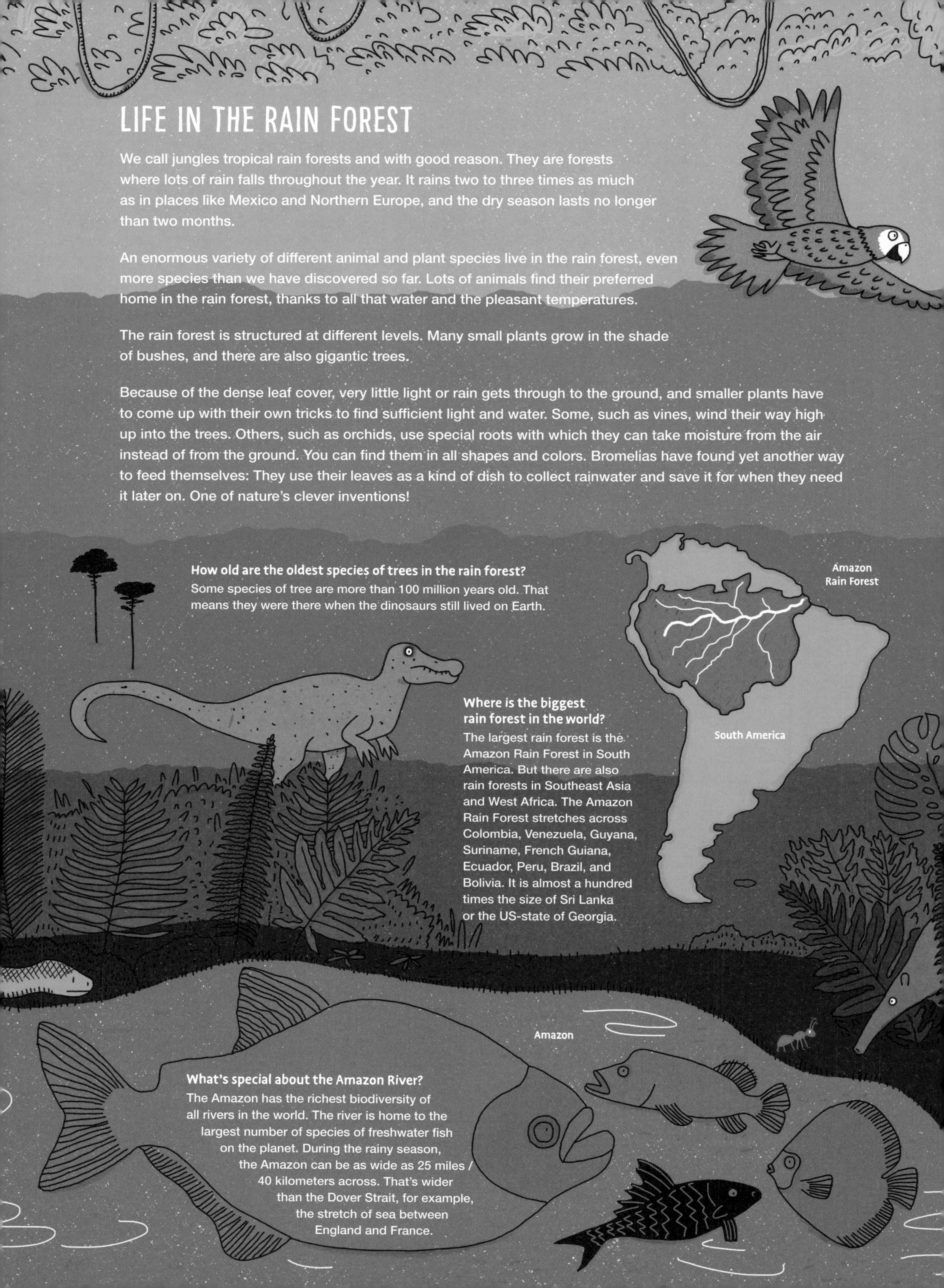

LIFE IN THE RAIN FOREST

We call jungles tropical rain forests and with good reason. They are forests where lots of rain falls throughout the year. It rains two to three times as much as in places like Mexico and Northern Europe, and the dry season lasts no longer than two months.

An enormous variety of different animal and plant species live in the rain forest, even more species than we have discovered so far. Lots of animals find their preferred home in the rain forest, thanks to all that water and the pleasant temperatures.

The rain forest is structured at different levels. Many small plants grow in the shade of bushes, and there are also gigantic trees.

Because of the dense leaf cover, very little light or rain gets through to the ground, and smaller plants have to come up with their own tricks to find sufficient light and water. Some, such as vines, wind their way high up into the trees. Others, such as orchids, use special roots with which they can take moisture from the air instead of from the ground. You can find them in all shapes and colors. Bromelias have found yet another way to feed themselves: They use their leaves as a kind of dish to collect rainwater and save it for when they need it later on. One of nature's clever inventions!

How old are the oldest species of trees in the rain forest?
Some species of tree are more than 100 million years old. That means they were there when the dinosaurs still lived on Earth.

Where is the biggest rain forest in the world?
The largest rain forest is the Amazon Rain Forest in South America. But there are also rain forests in Southeast Asia and West Africa. The Amazon Rain Forest stretches across Colombia, Venezuela, Guyana, Suriname, French Guiana, Ecuador, Peru, Brazil, and Bolivia. It is almost a hundred times the size of Sri Lanka or the US-state of Georgia.

What's special about the Amazon River?
The Amazon has the richest biodiversity of all rivers in the world. The river is home to the largest number of species of freshwater fish on the planet. During the rainy season, the Amazon can be as wide as 25 miles / 40 kilometers across. That's wider than the Dover Strait, for example, the stretch of sea between England and France.

vines
bromelia
orchid
water lily

SURPRISES IN THE DESERT: OASES

Picture yourself in the desert. The sun is high in the sky, and the heat is blazing. There's nothing around you but sand, sand, and more sand. Then suddenly you see something: some date palms and green bushes, a few huts, signs of life. It's an oasis!

An oasis is an area of land where plants can grow in the middle of a very dry region because there is freshwater. Sometimes it's a natural spring. However, there are also a great many oases where humans have given nature a helping hand, channeling water through underground aqueducts or pumping it in. A famous example of underground aqueducts that transport water to the surface are the qanats found in the Middle East. These are also found in North Africa and Central and Western Asia.

Oases have been very important to humankind for centuries. Anyone who traveled across the desert to sell goods in a town on the other side was able to rest there. For hundreds of years, people have been planting hardy date palms to protect their oases against wind and sand.

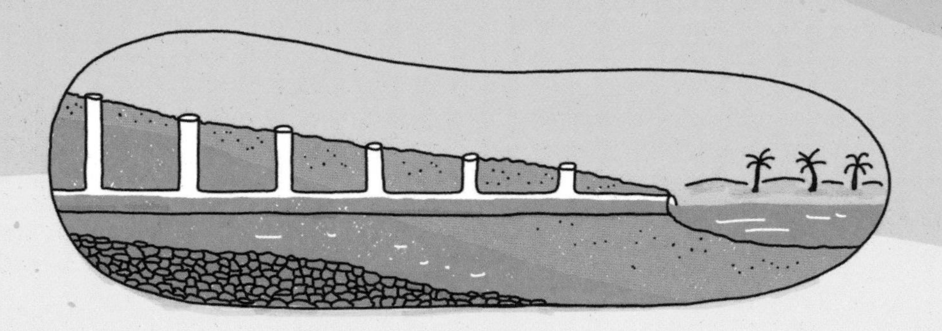

How do qanats work?

Qanats are underground aquifers that transport water from springs in the mountains to the lower-lying, drier areas. The technology has been used for hundreds of years. From a valley, humans dug out a large tunnel for dozens of miles / kilometers to reach the spring. On the way, the people doing the digging also created vertical shafts to allow air into the tunnel. Later, once the qanat was ready to work, they also used the shafts to keep the qanat clean. Ancient qanats are still in use today because they rely on gravity to take water to precisely those places where it's needed.

How does a camel survive in the desert?

You might think the camel's humps contain water, but they don't. They mainly contain fat, which serves a reserve for when the camel has to go for a long time without food, which can sometimes happen in the desert. But how do camels manage to survive so long without water? The key is their kidneys, the organs that make urine by separating nutrients from waste products. The nutrients are absorbed, and the waste products are passed out. A camel's kidneys barely need any moisture, so as a result, a camel's urine, unlike that of humans, comes out as a kind of thick syrup.

Where does the water in the oasis come from?

Although it's very dry in the desert and there is barely any water to be found, there are still a few places where water naturally comes to the surface. We call such places natural water sources.

What is a fata morgana?

A fata morgana is a mirage. When different layers of air at very different temperatures meet, it causes the air to behave like a mirror. For instance, it can look as if a ship is reflected in the distance, but we also use the term "fata morgana" to indicate that we are seeing an illusion, something that's not real. In comics and cartoons, a fata morgana is often portrayed as an oasis with an abundance of food and drink.

Which famous city was built at an oasis?

The popular casino city of Las Vegas in the United States is right by a natural oasis in the Mojave Desert. Las Vegas means "the meadows" in Spanish. This was the name the Mexican traders gave the place when they discovered it in 1829.

WHAT IS TIGER BUSH?

In places where it doesn't rain much, such as in the Sahel in Africa or the Baja California peninsula in Mexico, you can find a special sort of forest: the tiger bush. Bare ground alternates with stretches of trees, bushes, or grass that follow the contour lines. Contour lines are lines on a map connecting places that are at the same height level, or altitude. If you were to follow a contour line on a mountain, you would never walk uphill or downhill.

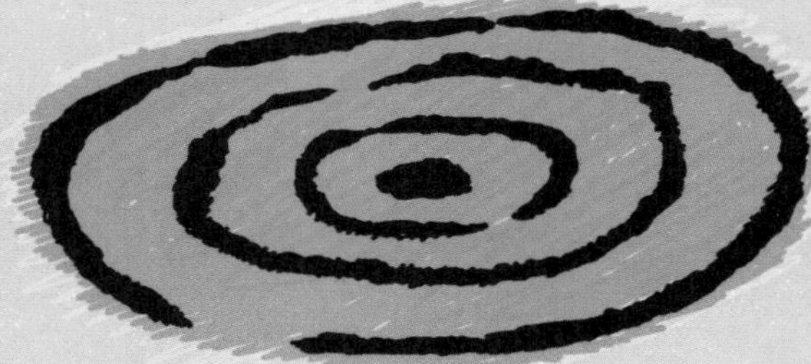

overhead view

But do you know why this tiger bush grows in stripes like the fur of a tiger and doesn't grow all over or in random clumps? Or how the trees and bushes in the tiger bush manage to survive at all in such dry areas? Many of those trees and bushes normally grow only in areas where it rains more frequently.

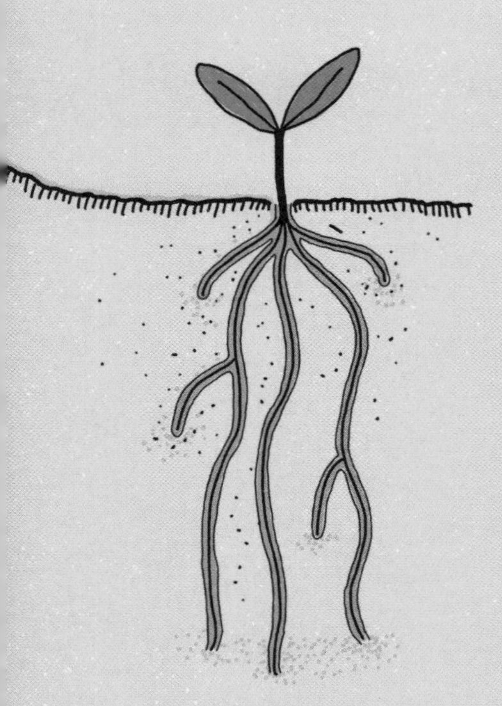

The root pushes soil particles aside and makes its way downward. This creates channels and makes the soil looser.

If there is little or no rain, plants have to get their water from elsewhere. Their roots have to search wide and deep in the ground to find water. But if a plant succeeds in surviving this way, its roots and falling leaves improve the soil. More gaps and channels are formed, through which the water can flow and be stored when it does rain. This helps the plant grow better in that particular place.

When rain is scarce, the top layer of the soil becomes very dry and hard. It even slightly repels the first rainwater. The water that doesn't immediately penetrate the ground then flows over its surface and runs off. Where the plants grow, the soil has more gaps and little channels, so it can better absorb the water when it rains. This is very good for the plants growing there. They not only take up the rain falling on their own spot of ground but also the water flowing down from slightly higher ground.

When you look at tiger bush from the air, it looks like the magnificent coat of a tiger—stripes of dark tiger bush over yellowy-orange dry ground. And this is exactly how it got its name.

LET'S DO IT!

≈ Take some bone-dry potting soil and put it into two bowls.
≈ In one bowl, mix some water with the soil. Don't add so much that the water pools on top of the soil.
≈ Press the soil down in both bowls.
≈ Then pour a few drops of water onto the soil in each bowl. What happens?

Answer: The water doesn't immediately sink into the dry soil. It stays on top. On the damp soil, the water sinks in right away.

Do tigers live in the tiger bush?

Tiger bush is important for grazing animals, such as giraffes. These animals live off the leaves growing on the bushes. You probably won't come across any tigers here, however.

How old is tiger bush?

Tiger bush seems to always have existed, yet even a small change in the amount of rainfall can bring about separate groups of bushes or make an ordinary forest appear in less than ten years. That's because small changes can suddenly mean that hardly any water flows down anymore, so there is no benefit to plants that try to reach it along the contour lines. The opposite can also happen: If a little more water falls, there is suddenly sufficient water for the plants everywhere and they no longer need to gather in stripes.

RIVERS AND FLOODS

Water runs down to the sea through small, babbling streams and big, wide rivers, but where do you think the water in those watercourses comes from?

Some of the river water is rain that has fallen on the land. The raindrops flow across the fields and streets. The water flows from high to low, until it ends up in a stream or river. Meltwater from snow and glaciers also ends up in rivers.

Much of the water in rivers is groundwater. That sounds odd because didn't we say that groundwater is under the ground? That's true, but the water underground doesn't stay still either. It flows very slowly toward the streams, rivers, and seas. The treated sewage water also ends up in the rivers.

Sometimes it rains so much that the water in a river rises very rapidly. The river overflows! Why is that? Well, people like to straighten the course of rivers so that they can use the land right up to the river's edge. They want to build houses or grow crops on it. Yet rivers would overflow far less often if we allowed them to follow their natural, winding courses and let them flood the meadows.

When the river floods, fields, roads, and houses can end up underwater. You've probably seen it on TV! In extreme floods, people may even be swept away or are left stuck on the roofs of their cars or houses.

What does a dike do?

Dikes were built to protect people and their houses on the land from flooding—for example, the levees by the Mississippi River or the dikes in the Netherlands. Dikes can break when the force of the water is too strong, allowing the water to flood the lower-lying land. After major floods in 1953, the Dutch dikes had to be built to a greater height. Since then, most floods in the Netherlands have happened near the rivers and not by the sea.

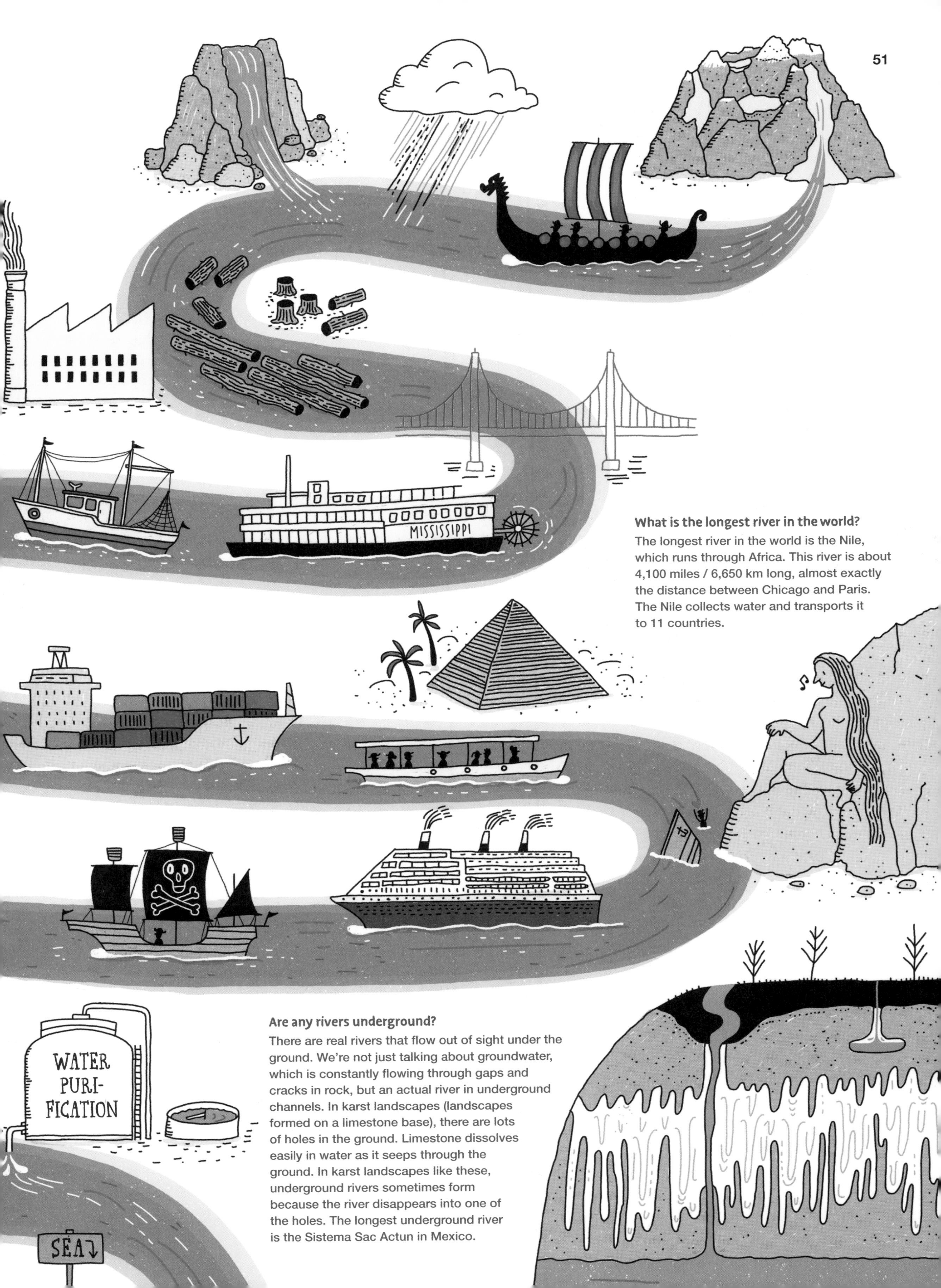

What is the longest river in the world?

The longest river in the world is the Nile, which runs through Africa. This river is about 4,100 miles / 6,650 km long, almost exactly the distance between Chicago and Paris. The Nile collects water and transports it to 11 countries.

Are any rivers underground?

There are real rivers that flow out of sight under the ground. We're not just talking about groundwater, which is constantly flowing through gaps and cracks in rock, but an actual river in underground channels. In karst landscapes (landscapes formed on a limestone base), there are lots of holes in the ground. Limestone dissolves easily in water as it seeps through the ground. In karst landscapes like these, underground rivers sometimes form because the river disappears into one of the holes. The longest underground river is the Sistema Sac Actun in Mexico.

WAVES AND  TSUNAMIS

You can't beat jumping over the waves on the beach in summer! The waves form when there are high winds at sea. The wind pushes the top layer of water upward, and the wave grows bigger and bigger. In heavy storms, the waves can become very high indeed.

There is another cause of high waves known as tsunamis. These happen after an earthquake or volcanic eruption at sea. When the earth shakes, it can result in an enormous wave. The word *tsunami* combines two Japanese words: *tsu* ("harbor") and *nami* ("wave"), together making "harbor wave." A "wall" of water moves toward the land. It can be as tall as a house or even an apartment building by the time it reaches the coast! A tsunami has immense force and destroys everything in its path. It can travel at ten times the speed of a car on the highway.

The only thing you can do if a tsunami is approaching is get to a higher place as fast as possible, away from the coast. To make sure that people can get to safe places in time, many countries have tsunami alarms. When an earthquake happens at sea, the alarm goes off and people know that they need to find a safe place quickly.

Why is the North Sea around Belgium and the Netherlands brown?
The water of the North Sea is brown because there are lots of small particles floating in it, such as sand, clay, algae, and bacteria. In our coastal waters, there are 0.01 to 0.3 ounces per cubic foot / 10 to 300 milligrams per liter of those particles in the water, sometimes even more if high waves stir up the silt during a storm. The silt particles come from the French and Belgian coasts and from rivers such as the Scheldt, Maas, and Rhine. In the coastal water, the silt particles are scattered farther. In calm weather, they sink back down to the seabed, but they are stirred up again by strong tidal currents and waves in the North Sea.

The wind forms waves.

An earthquake or an undersea volcanic eruption can cause a tsunami.

How do waves form in a swimming pool?
Some water park swimming pools have waves. These waves are not caused by the wind. They are created by large fans set up around the swimming pool. The air from the fans pushes the water into waves. Different types of waves can be selected by using valves.

Why is there sometimes foam on the waves in the sea?
Very small marine plants called algae live in the sea. When the algae die, they leave a sticky substance behind in the sea. Then the waves turn this into foam, or spume. The waves and current wash the foam onto the beach.

Where do tsunamis happen?
Tsunamis do not happen everywhere equally often. Some regions are at greater risk of an earthquake happening under the seabed. This is the case in the region around the Pacific Ocean, and that's why more tsunamis happen there.

DAMS AND RESERVOIRS

Some rivers have one or more dams built on them. It's a kind of wall that holds back the water so that a lake forms behind the dam. We call this lake a reservoir. The water from the reservoir can be used to make drinking water or to irrigate fields.

A dam can also help us produce electricity. The water flows from the high-up reservoir over a kind of wheel in the lower-lying river, turning the wheel on its way. The turning force that the water transfers to the wheel can be used to make electricity. Energy that is produced using water is known as hydropower.

Dams have their disadvantages too. The land to one side of the dam ends up underwater, making fields and sometimes entire villages disappear. The river on the other side of the dam often receives much less water than before. This can be a problem for the people who use the river water and for the animals who live in it. It is the reason local people often protest when a new dam is planned.

Sometimes accidents happen with dams. If too much water gathers in the reservoir behind the dam, the structure can break. The water then rushes down with great force and can cause a lot of damage and destroy entire villages. The biggest dam disaster happened in China in 1975, when 26,000 people died after the Banqiao Dam collapsed.

Which dams are the most famous in the world?

The longest dam in the world is the Hirakud Dam in India. It is 16.5 miles / 27 kilometers long. There are many famous dams, such as the Hoover Dam in the United States, the Aswan Dam in Egypt, and the Atatürk Dam in Turkey. The Three Gorges Dam in China produces the most electricity.

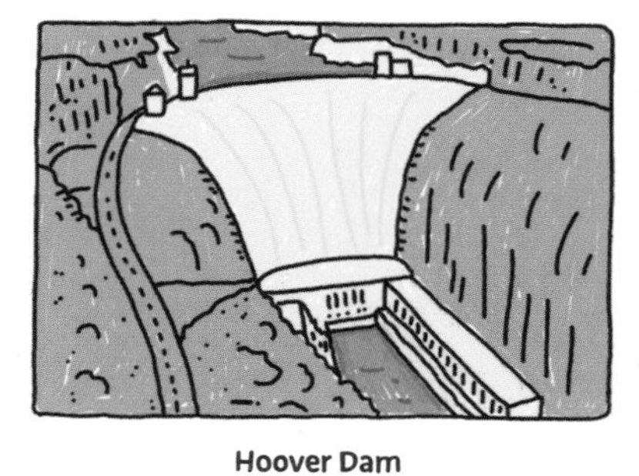

Hoover Dam

Atatürk Dam

Banqiao Dam

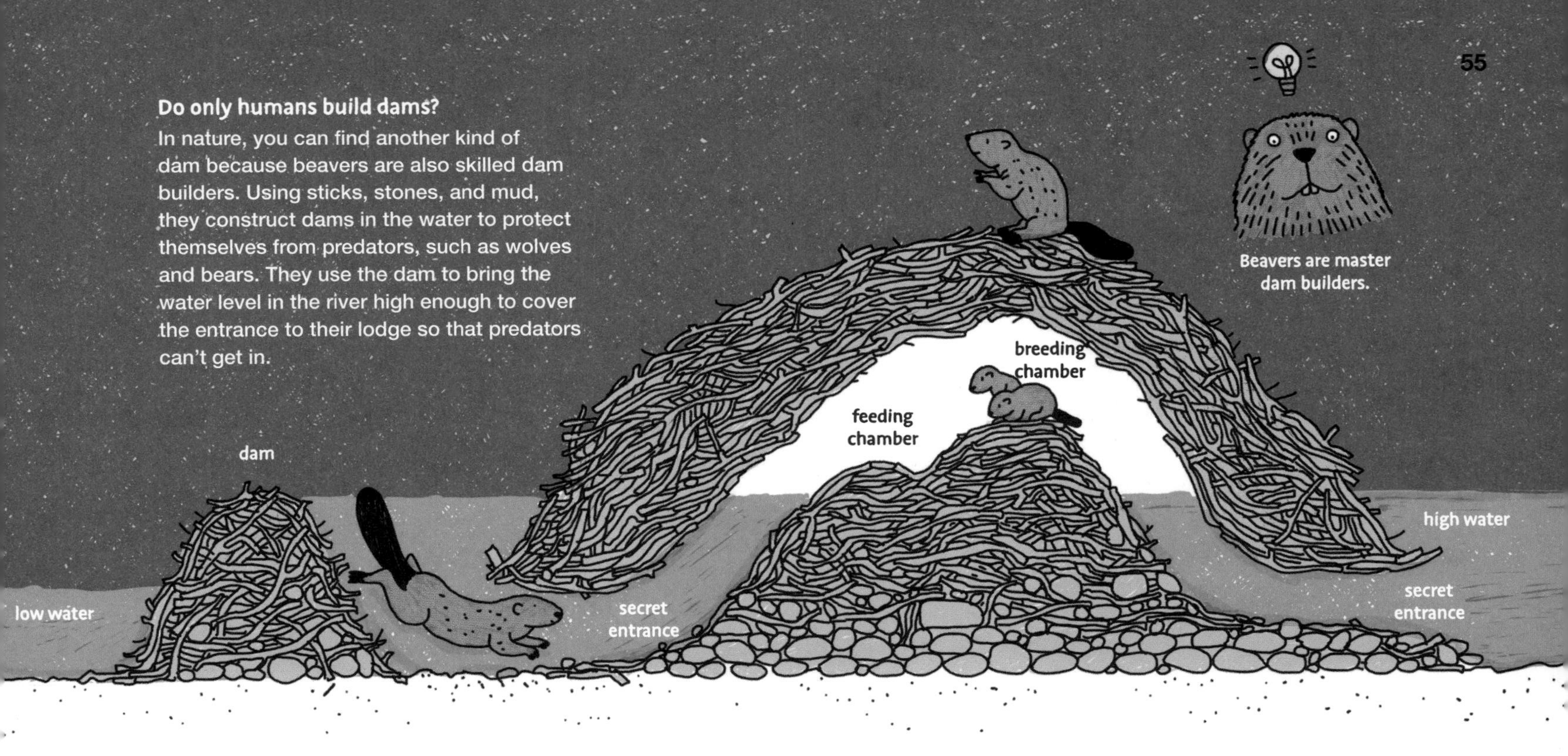

Do only humans build dams?

In nature, you can find another kind of dam because beavers are also skilled dam builders. Using sticks, stones, and mud, they construct dams in the water to protect themselves from predators, such as wolves and bears. They use the dam to bring the water level in the river high enough to cover the entrance to their lodge so that predators can't get in.

What is a flood barrier?

A flood barrier is a kind of long dam that separates two areas and ensures that the lower-lying land is not flooded when the water levels are high. Special barriers rise with the water, so even in the case of extremely high water, the land behind the barrier is protected. The old Dutch fishing village of Spakenburg is entirely surrounded by a flexible barrier of this kind that is 360 yards / 330 meters in length. The inventor of this barrier was the Dutch artist Johann van den Noort, who thought up this system when his own village, Kampen, was threatened with flooding. The flexible barrier is now used all over the world.

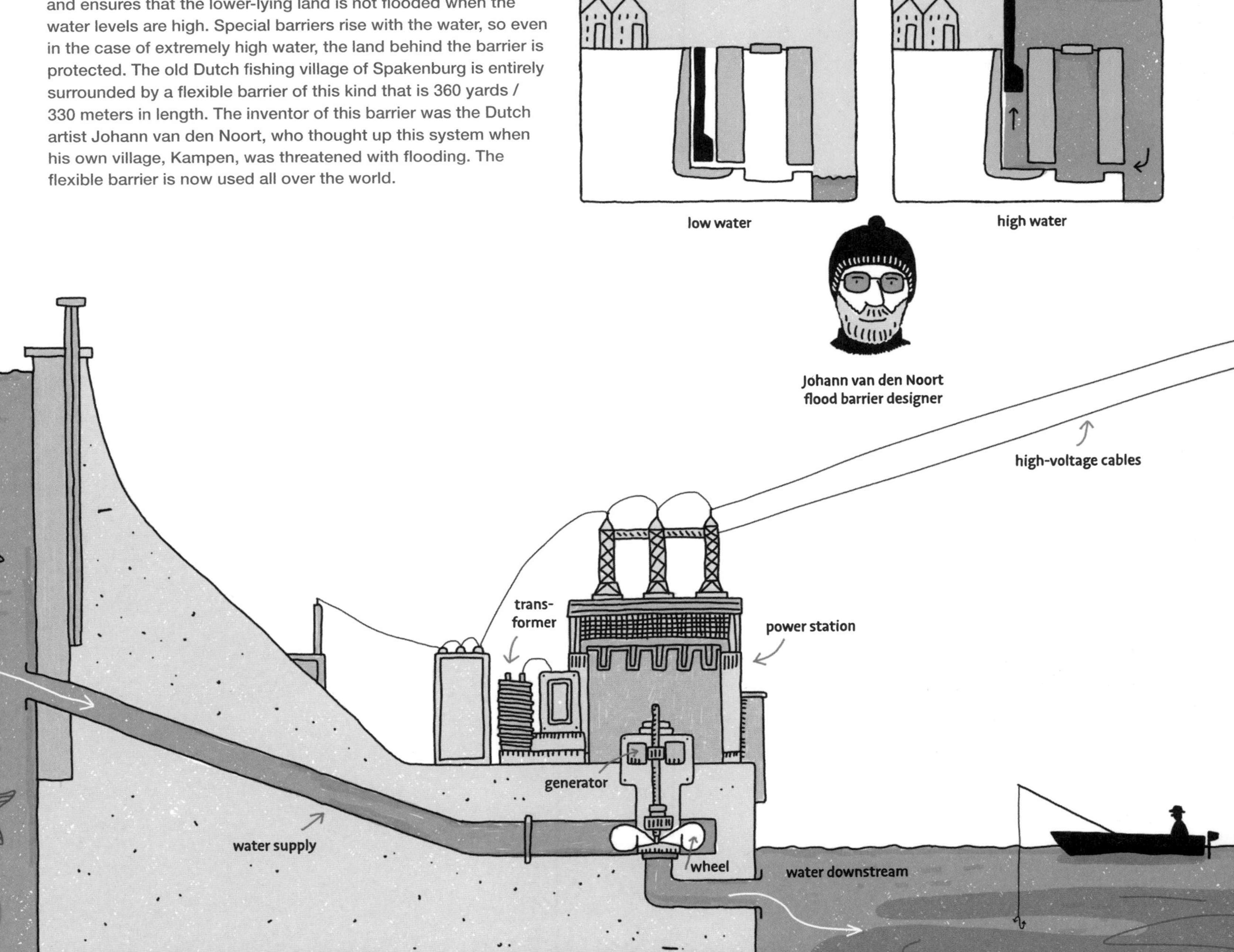

THE FUTURE OF WATER IN THE WORLD

Our climate is changing. The planet Earth is growing warmer because people release too many greenhouse gases into the air by burning coal, oil, and gas. The heat is melting a great deal of ice and snow. That meltwater runs into the sea, raising the sea level. If we don't take action, many places where people currently live and work will be covered by the sea. This is a particularly serious challenge in low-lying places like the East Coast of the United States, including New York City and Miami, major cities in South East Asia, such as Bangkok and Shanghai, and the Netherlands, where dikes are already needed to protect the land from the sea!

Climate change also has consequences for water on the land. Some places are becoming wetter and others dryer. Sometimes that's good news. For instance, there are deserts that are becoming wetter due to climate change. In most places on Earth, however, climate change is bad news. Often wet regions become wetter, with even more floods, and dry regions become so dry that there is even less water to grow food.

In some countries, it is becoming both wetter and dryer depending on the season, so we need to expect such places to suffer greater climate extremes, with more floods and longer periods of drought.

It is not only the climate that determines how much water we will have in the future. In many places in the world, humans themselves cause water shortages. For example, in some places in China, India, and the United States, a lot more groundwater is pumped out of the ground than nature can top up. The water supply is dropping very quickly in those places. Within a few years, the groundwater could run out altogether, leaving too little water to produce food.

Why is polluted groundwater a major problem?
Since groundwater moves very slowly underground, pollution from humans stays in the water for a long time. It can come from different sources—for example, farming or factories. With current technology, we don't generally succeed in making polluted groundwater completely clean, so the pollution will remain in the water for many generations. Therefore, it is important to ensure that we don't pollute even more groundwater and that we purify our wastewater properly before returning it to nature.

Why should we worry about melting glaciers?

In the world's mountainous regions, such as the Himalayas and the Andes, a lot of water is stored in glaciers and snow. The meltwater from these glaciers feeds the rivers, and these supply people with water. Unfortunately, the glaciers are now melting rapidly. If they disappear altogether, people will no longer have the same quantity of river water that they had before.

How can we predict the climate?

When we talk about climate change, we really want to predict the weather in a particular place for the next hundred years. That's not a simple task. Even predicting next week's weather is not easy. The weather forecasters, unfortunately, often get it wrong. We have developed a number of different methods for calculating future weather. As in the saying "There's safety in numbers," we combine all these calculation models in order to predict the climate. In this way, we can see where all the calculations agree and where they come to different conclusions, so we are clear about the certainty or uncertainty of our predictions. Although the magnitude of climate change is not always exact, it is clear that humans are causing never-seen-before changes to the climate.

HOW CAN YOU SAVE WATER?

Take a short shower instead of a bath, or share a bath.

Make sure your feet are clean before you enter an outdoor swimming pool!

Use the bath or shower water to flush your toilet. You can do it with a bucket.

Try to buy secondhand clothes instead of new ones.

Do you need to water the plants in your yard? Use a watering can instead of a hose. That way, you water the plants without wasting too much water.

If you have to wait a moment for the water to become warm, collect the cold water and use it for your plants.

A lot of water is needed to produce meat. Eat meatless dishes from time to time.

Repair leaking faucets or pipes straightaway.

Keep the water you use to wash vegetables or for other kitchen tasks and give it to the plants in the yard.

Collect rainwater and use it for the toilet, washing machine, and yard. Do not use it as drinking water.

You don't need to water grass when it's dry. Let it go yellow or brown. It will turn green again the next time it rains.

Choose water-saving household appliances: eco-showerheads, faucets, dishwashers, and so on.

Use the light toilet flush whenever you can.

Use the same glass all day to save on washing up.

Turn the faucet off while brushing your teeth.

DICTIONARY

Air pressure ≈ p. 14
An air-pressure measure tells us how much force air uses to push down on us and everything around us. That force gets smaller and smaller the higher up you go in the air.

Climate change ≈ p. 38, p. 56
The climate can change. The change can be caused by changes in nature itself or in human activity. Humans are changing the climate by burning oil and gas, which releases surplus carbon dioxide into the air. This surplus starts off a chain reaction that causes Earth to heat up.

Condensation ≈ p. 15
The transformation of water vapor into liquid water.

Evaporation ≈ p. 15
The transformation of liquid water into water vapor.

Fata morgana ≈ p. 47
When different layers of air have different temperatures, the air can act like a mirror. Normally the air itself is reflected, making it look like there is a pool of water in the distance but, in fact, that's only the reflected sky.

Freezing ≈ p. 15
The transformation of liquid water into solid water (ice).

Geology ≈ p. 22
The scientific discipline that studies the structure of the planet Earth and how it changes, as well as its natural and mineral resources.

Glacier ≈ p. 57
When a thick layer of snow stays on the land for a long time without melting, the snow can turn into ice. An ice mass of that kind on land is known as a glacier.

Infiltration ≈ p. 16
We call the process of rainwater disappearing into the ground infiltration.

Irrigation ≈ p. 18, p. 38
When it's dry for a long time, farmers cannot rely on rainfall. They have to water the plants themselves. We call this irrigation, and it can be done in various ways: by flooding a field, by spraying with sprinklers, or with the help of irrigation tubes.

Melting ≈ p. 15
The transformation of solid water (ice) into liquid water.

Photosynthesis ≈ p. 42
Photosynthesis is the process whereby plants, in the presence of sunlight, convert water (H_2O) and carbon dioxide (CO_2) into oxygen (O_2) and food (sugars).

Rain shadow ≈ p. 25
A rain shadow is a phenomenon that often occurs in mountainous areas. Moist air is forced to rise against the windward side of the mountain and rains down, partially or completely. The windward side is the side of the mountain against which the wind blows. Meanwhile, on its other side, the leeside, the mountain remains dry; it is located in the rain shadow.

Reservoir ≈ p. 29, p. 32, p. 54
In some rivers, there is a dam, a kind of wall that holds back water. Upriver behind the dam, a lake forms, and this is known as a reservoir.

Sublimation ≈ p. 15
The transformation of solid water (ice) into water vapor, without the water first becoming liquid.

Tides ≈ p. 13
The sea sometimes comes closer and sometimes moves farther away down the beach. Those movements are known as tides. Seas and oceans move because they are attracted by the moon, which rotates around Earth every day. When the water rises, that's known as high tide, and when the water goes down, we call it low tide.

The sun also pulls on Earth's water a bit. When the sun and the moon pull in the same direction, we have spring tides and the water can rise very high.

Transpiration ≈ p. 16
The water that plants take from the ground through their roots comes back out into the air through their leaves. We call that transpiration.

Virtual water ≈ p. 40
We call all the water that is needed to make a particular product and to compensate for pollution caused by this process virtual water.

Water footprint ≈ p. 40
The water footprint of a product is the sum of all the water needed to make that product. It's not just about the water that is used up; the water that is polluted in the process also makes up part of the water footprint.

Sarah Garré is a researcher at ILVO (the Institute for Agricultural and Fisheries Research) and professor at the University of Liège. She is passionate about the role of water on our planet and how we can manage it better.

Marijke Huysmans is a professor at Vrije Universiteit Brussel and KU Leuven. She is fascinated by the water that is in the ground beneath our feet.

Wendy Panders is an illustrator and graphic designer. She enjoys illustrating stories and informative books for children.

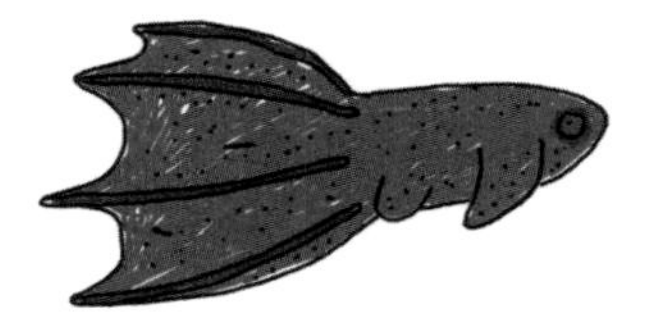

Original title: Het grote waterboek. Van zwetende planten tot verwoestende tsunami's.
Translated from the Dutch language
www.lannoo.com

A member of Penguin Random House Verlagsgruppe GmbH
Neumarkter Strasse 28 · 81673 Munich

Library of Congress Control Number: 2022945718
A CIP catalogue record for this book is available from the British Library.

Nederlands
letterenfonds
dutch foundation
for literature

The publisher gratefully acknowledges the support of the Dutch Foundation for Literature.

Translated from the Dutch by Anna Asbury

Project management: Constanze Holler
Copyediting: Marilyn Knowlton
Production management: Susanne Hermann
Typesetting: Sylvia Goulding
Printing and binding: Neografia a.s., Slovakia

Prestel Publishing compensates the CO_2 emissions produced from the making of this book by supporting a reforestation project in Brazil. Find further information on the project here:
www.ClimatePartner.com/14044-1912-1001

Penguin Random House Verlagsgruppe FSC® N001967

Printed in Slovakia

ISBN 978-3-7913-7550–2

www.prestel.com